Medicating Modern America

Medicating Modern America

Prescription Drugs in History

EDITED BY

Andrea Tone and Elizabeth Siegel Watkins

New York University Press
NEW YORK AND LONDON

NEW YORK UNIVERSITY PRESS
New York and London
www.nyupress.org
© 2007 by New York University
All rights reserved

Library of Congress Cataloging-in-Publication Data
Medicating modern America : prescription drugs in history / edited by
Andrea Tone and Elizabeth Siegel Watkins.
p. cm.
Includes bibliographical references and index.
ISBN-13: 978-0-8147-8300-9 (cloth : alk. paper)
ISBN-10: 0-8147-8300-7 (cloth : alk. paper)
ISBN-13: 978-0-8147-8301-6 (pbk. : alk. paper)
ISBN-10: 0-8147-8301-5 (pbk. : alk. paper)
1. Drugs—United States—History. I. Watkins, Elizabeth Siegel.
RS67.U6M43 2006
615'.1—dc22 2006022746

Contents

Introduction

Medicating Modern America explores the rich and multifaceted history of pharmaceutical medicines in modern America since World War II. With Americans paying more than $200 billion in 2005 for prescription pills, the pharmaceutical business is the most profitable in the nation.[1] The popularity of prescription drugs in recent decades has reframed interactions between doctors and patients, making prescription-writing and pill-taking an integral part of medical practice and everyday life. *Medicating Modern America* examines the stories and meanings behind this pharmaceutical revolution through the discrete but interconnected histories of some of the most influential and important drugs associated with its rise.

Drugs are substances that alter the body in order to alleviate symptoms, help make a diagnosis, or promote health and well-being. Their use dates back to ancient times. In cultures and communities around the globe, people have turned to medicinal preparations of plant, animal, or mineral origin to ease physical and psychic pain, combat infection, induce sleep, or stop the spread of disease. Quinine, belladonna, opium, cannabis, and alcohol are but some of the many modern substances that have been in pharmacological circulation for centuries.[2] Our focus here, however, is on a more recent category of therapeutic agents: doctor-prescribed medicines whose surging popularity in the last half-century marked the advent of what some have called the "golden age" of pharmaceutical science. The authors in this volume take an explicitly historical approach to studying the development, prescription, and consumption of these drugs. This perspective locates the histories of prescription medicines in specific socio-cultural contexts while revealing the extent to which contemporary debates about pharmaceutical drugs revisit concerns expressed by Americans over the past several decades.

Contemporaries rightly regarded the 1940s and 1950s as a new chapter in the history of modern medicine. The mass manufacture of orally acting antibiotics—first penicillin, then streptomycin and the tetracyclines—fanned widespread faith in the possibility of laboratory solutions to vexing human troubles. Magazine articles and children's books retold the story of Alexander Fleming's accidental discovery of the penicillin mold in 1928. Newsreels brought viewers onto the drug factory floor, with mesmerizing moving images of pills traveling down assembly lines under the watchful eyes of white-coated attendants. By 1944, pharmaceutical companies had pioneered new technologies that permitted the mass production of penicillin. By the end of the war, commercial production—which rose from 400 million units per month in the first half of 1943 to a staggering 650 billion units by August 1945—had saved thousands of soldiers' lives.[3] In civilian medicine, antibiotics offered doctors and patients their first real cure for syphilis, tuberculosis, bacterial pneumonia, and other infectious diseases. Penicillin became a symbol of what laboratory- and corporate-driven science could offer, and it was with good reason that Fleming, along with bacteriologists Howard Florey and Ernst Chain, shared the Nobel Prize for medicine in 1945.

Penicillin was followed by other pharmaceutical breakthroughs: corticosteroids, broad-spectrum antibiotics, tranquilizers, antidepressants, antipsychotics, and the first pills to treat hypertension and diabetes. Although several authors in this volume question the extent to which the new therapies provided alternatives better than what they had replaced, it would be hard to downplay contemporaries' own enthusiasm. Americans evaluated pharmaceutical medicines in a cultural and political context optimistic about the possibilities of Cold War science. The mass manufacture of penicillin and the insecticide DDT, whose discoverers also won the Nobel Prize in 1948, were powerful tributes to the success of institutional research. The new pharmaceuticals gave doctors a veritable armamentarium of drugs to treat a wide array of disorders. In the eyes of many, these agents were nothing less than "magic bullets": medicines that acted on specific diseases but left the patients who housed them intact.[4] Only a few decades earlier mercury, which in high doses could impair vision, induce kidney failure, and even kill, had been the treatment of choice for syphilis, and lobotomies, which irrevocably transformed patients' personalities as well as their brains, were considered a reasonable cure for psychosis. Remedies, in short, could seem as dangerous as the illnesses they were meant to help.

Against this backdrop, the availability of pharmaceutical drugs promised easier, seemingly less invasive, and in many cases, more effective therapies. Illnesses that had previously been regarded as medically insurmountable could now be controlled, even cured, by something as simple and streamlined as a pill. Treatments that had once confined patients to institutions with barred windows could now be administered on an outpatient basis. Mass expansion after the war of federally funded research in the National Institutes of Health and the National Science Foundation had led many Americans to believe that they would personally benefit from the scientific research being conducted in laboratories across the country. The arrival of a therapeutic regime grounded in drugs-by-design confirmed this vision. The dawn of the age of the wonder drug appeared to be nothing short of miraculous.

From the beginning, doctors and drug companies were handmaidens to the new pharmaceutical order. In 1951, Congress passed the Durham-Humphrey Amendments to the 1938 Food, Drug, and Cosmetic Act. This important law, a response to widespread concerns that risky drugs could be effortlessly purchased and harmfully consumed, set up the federal prescription-only classification, mandating that consumers of new and potentially dangerous drugs acquire them only after consulting a physician and obtaining a prescription. Before the 1950s, prescription status had applied only to narcotics and cocaine; unevenly enforced state laws left the rest of the drug market largely unfettered and unregulated.[5] The 1951 measure changed all of that, establishing doctors as the expert gatekeepers to drugs and positioning reputable and, increasingly, large-scale pharmaceutical companies as the primary suppliers of prescription medicines.[6]

The decades that followed cemented the power and prestige of pharmaceutical medicines. From antibiotics to Viagra, one of the threads of continuity in the postwar period has been American's resounding enthusiasm for prescription pills. This enthusiasm has endured within a political and social context that has often been critical of the country's drug intake. Prescription medications have spurred Congressional investigations and fomented the passage of new laws to restrict how drugs are manufactured, marketed, labeled, prescribed, sold, and used.

In the 1980s, the appearance of new diseases, such as AIDS, and the recurrence of old diseases, such as tuberculosis, brought further criticism of the drug industry. AIDS advocates contested the high cost of drugs such as AZT, and health care workers struggled to provide the long-term antibiotic therapy needed to cure tuberculosis. Drugs such as antidepressants and hormone replacement therapy galvanized activists and consumer groups, many of whom charged that drugs have been inappropriately prescribed for conditions that lifestyle changes could better handle. In the 2000s, the withdrawal of drugs such as the anti-obesity agent Fen Phenn (a combination of fenfluramine and phentermine) and the analgesic Vioxx (rofecoxib) following reports of adverse side-effects made headline news and raised concerns about the FDA's ability to protect consumers from dangerous medicines. They have also made pharmaceuticals a lightning rod for litigation, as patients and families looking to hold someone responsible for their suffering turn to the courts for compensation.

Despite broad concerns and specific drug scandals, Americans continue to consume more prescription drugs than citizens of any other country. In 2004, roughly half of all Americans used at least one prescription drug daily. This behavior has not only stoked the profits of "Big Pharma," but it has also transformed the economy of the United States and the world. In 2004, global pharmaceutical sales surpassed the $500 billion threshold for the first time. With fewer than 5 percent of the world's population, the United States accounts for almost 50 percent of global sales.[7]

Although pharmaceutical drugs have transfigured social and medical habits, their history has been understudied, especially relative to other fields. The essays in this volume complement a burgeoning interest in this critical chapter of modern medical history. Each addresses the history of a specific drug. Rather than simply chronicling that drug's rise, diffusion, or demise, however, authors look closely and critically at the meanings that can be discerned from the pharmaceuticalization of modern America.

The themes authors explore engage ongoing and fruitful subjects of critical inquiry. In the realm of pharmaceutical innovation, for instance, recent scholarship has upended the Whiggish idea of rational drug development, challenging the myth that great minds sifting through the results of carefully crafted experiments were able to create new compounds to heal humankind. Despite familiar tales of the neatly linear development of drugs, many have occurred accidentally and haphazardly, and knowledge about what has been discovered and what should be done with it has often followed circuitous and unpredictable paths. The complexity of medical and commercial trajectories is exemplified by the fate of the penicillin mold, whose therapeutic implications were not fully appreciated until 1939, eleven years after Fleming encountered it.[8] Often, drugs intended to treat one disorder have been medically used for an entirely different condition. Further undermining the notion of a rational timetable of drug innovation is the explosion of the so-called me-too drugs, copycat medications that modify only marginally the chemical structure of older drugs. In the United States, the majority of new drugs approved by the FDA in the last ten years cost more but offer "no significant clinical improvement" over older compounds.[9]

Historians have also elucidated the multiple political, commercial, and social routes by which drugs come to "meet" standards of therapeutic efficacy, by demonstrating that standards for measuring a drug's efficacy and safety have been constructed and contested in different realms. Although large-scale randomized controlled trials are held up as the emblem of objective knowledge, the protocols and results of clinical trials may be best understood as negotiations among a wide set of social actors—regulators, patients, physicians, activists, and pharmaceutical executives—each of whom is differently invested in what gets counted as a "medical fact."[10] Recent work on the influence of the pharmaceutical industry has underscored the power of business to decide not only the kinds of drugs and disorders that warrant financial support but also which trial results are worth sharing with the Food and Drug Administration.[11] This research raises important and often unsettling questions about who should have the power to make medicines and what criteria should be applied, and by whom, to determine when a drug can be considered safe and beneficial.

Related to these issues is the thorny matter of the determination of what counts as a legitimate drug. Americans schooled in the politics of "just say no" have divided the drug market into artificial realms of vice and virtue. A drug's value is measured not only by its pharmacological effects but also by the cultural contexts in which it is made, circulated, and used. On the one hand, many Americans express horror at the alleged epidemic of "bad drugs" such as heroin, methamphetamine, marijuana, and crack-cocaine, which are linked in movies, television shows, and even public policies, to street junkies, youth gangs, and persons of color. Because this drug trade has been branded illicit, the federal government spends millions of dollars trying to stop it. On the other hand, scientific evidence demonstrates that substances such as marijuana, for instance, are less addictive than alcohol, caffeine, or nicotine, all of which are sold legally.[12] Waging war on illegitimate drugs, political parties simultaneously accept millions from "legitimate" pharmaceutical houses. Between 1996 and 2002, for example, the pharmaceutical industry spent half a billion dollars on lobbying alone. In Washington, D.C., full-time lobbyists for the drug industry currently outnumber senators six to one.[13] Not all prescription medications have been regarded as equally licit and benign, of course;

often the public image of a particular drug has changed over time, in response to both scientific research findings and sensational media stories. This shifting reception reminds us of the power of political, cultural, and economic concerns to frame drug debates. The same social prejudices that have informed policies toward narcotics, a class of drugs whose history and stigmatization have been well studied, have colored attitudes toward other agents too.[14] More than potent chemicals, prescription drugs are social and political objects whose histories tell us much about the society and communities in which they are discussed, evaluated, and used.[15]

Historians and other scholars have also challenged the universality and fixity of the diseases and disorders that prescription drugs are meant to treat. To what extent are diseases and therapies socially or commercially constructed? The case of psychostimulants to treat Attention Deficit and Hyperactivity Disorder (ADHD) is instructive. Does the popularity of drugs such as Ritalin—invented in the mid-1950s but used on patients with attention deficit disorders only in the late 1960s—represent a pharmaceutical solution to a fixed physiological disorder, a decreased tolerance among American parents and educators for the natural but unpredictable rhythms of childhood and adolescent behavior, or something in-between?[16] By the same token, do women consume about twice as many antidepressants and anxiolytics than men because they suffer from depression or anxiety twice as much? Or is a female prevalence for psychotropic pill popping indicative of gendered social and economic patterns? After all, women are more likely than men to consult doctors, discuss symptoms in psychological terms, and buy medical products, including prescription drugs.

Determining the role of patients and consumers in the creation of the history of prescription drugs is also important. Decades ago, the historian Roy Porter encouraged scholars to pay heed to the economic power patients wield in medical markets. The paying consumer, he noted, "simply by possessing choice and the power of the purse, [can] exercise considerable sway in the medical marketplace."[17] This observation seems all the more apt when we explore the history of prescription drugs. Patients who "take their medicines" must take several steps: they must make a medical appointment, find time to see a doctor, get a prescription, go to a drugstore, purchase the drug, and then swallow the medicine itself. The tendency of patients to "do their own thing" with pills has often frustrated researchers and doctors and led to the creation of specially designed pill packages and elaborate systems of surveillance (such as directly observed therapy for tuberculosis) to promote patient compliance.[18] Yet, as Porter suggested, it is precisely the autonomous behavior that has consistently frustrated researchers to which scholars must pay attention if they wish to understand the complexity of patient and consumer involvement in the creation of America's prescription drug culture.

At the same time, patient choice in the prescription drug market cannot be understood outside of the many variables that circumscribe it, including a patient's financial means, doctors' prescription decisions, and the range of drugs available to patients in different locales. In recent years, scholars have looked critically at how the promotion and marketing of prescription drugs—first to doctors and more recently to consumers—have skewed our ideas about disorders, diseases, and the drugs we need to treat them.[19] Pharmaceutical companies defend advertising campaigns as important educational vehicles for teaching doctors and patients about medical problems and the availability of new therapies. Others cite advertising's social, economic, and medical costs. Drug promotion further medicalizes problems that were once the realm of the everyday, discussed and addressed outside of medical jurisdiction. Through million-dollar and carefully choreographed campaigns, companies have recast shyness, baldness, sexual dysfunction and other complaints into medical disorders severe enough to warrant pharmacological treatment. Many critics worry that pharmaceutical companies are redefining medical risk itself, in ways that blur the boundaries of when health ends and disease begins. Certainly advertising promotes the individualization of medical problems while proffering pills as a quick but ultimately patient-centered fix.[20] Because direct-to-consumer advertising is relatively new to the development of pharmaceutical culture, it may take some time before we can assess its long-term impact on people's attitudes about and uses of prescription drugs.

The authors in this volume have been part of this broader intellectual conversation; their work exemplifies current scholarship on the many-sided histories of twentieth-century pharmaceuticals in the United States. The eight drugs discussed here—antibiotics, mood stabilizers, hormone replacement therapy, oral contraceptives, tranquilizers, stimulants, statins, and Viagra—are by no means the only major pharmaceuticals to play a role in medicating modern America; rather, they stand as a revealing sample. We hope that the studies in this volume encourage other scholars to tackle issues presented by other drugs.

Although the eight drug classes in this book proved influential in the second half of the twentieth century, three of them (antibiotics, mood stabilizers, and hormone replacement therapy) trace their origins to the years prior to World War II. One—the antibiotics—fights germs; the others either modify mental states or prevent certain conditions from occurring. In his chapter on antibiotics, Robert Bud reflects on changing attitudes toward germs in the six decades since antibiotics came onto the market. He shows how a widespread obsession with the avoidance of germs gave way to a sense of invulnerability, a triumphal mood that was then eclipsed by fears of antibiotic resistance. He introduces the concept of branding to describe the various social meanings attached to antibiotics by pharmaceutical manufacturers, doctors, patients, and the media, as these drugs and their uses evolved in the context of the expanding consumerism of postwar America. Bud's chapter also reminds us that antibiotic use has been international; both the usages and meanings of antibiotics have changed not only across time but also by place.

The story David Healy tells also has an international component, chronicling how medical researchers in several countries over the course of the twentieth century coconstructed the development of mood stabilizer drugs and the classification of the disease category known today as bipolar disorder. However, he describes the increasing popularity of the diagnosis of bipolar disorder and the prescription of mood stabilizers as a uniquely American phenomenon. The American context plays an important part in Elizabeth Siegel Watkins's analysis of the changes in the production and consumption of information and advice about hormone replacement therapy over the past several decades. Watkins argues that social, cultural, and economic forces in the United States, as well as the evolving state of medical knowledge, shaped the development, delivery, and reception of information on estrogen and progestin therapies for postmenopausal women.

Three drug groups (oral contraceptives, stimulants, and tranquilizers) were products of the 1950s although, as each author deftly shows, medical and popular opinions about these drugs underwent significant changes throughout the century. The oral contraceptives that are the subject of the chapter by Suzanne White Junod are closely related to the hormone replacement therapy studied by Watkins, as both are composed of the hormones estrogen and progestin and are intended solely for women. Junod investigates how the risks associated with oral contraceptives were communicated to female consumers, highlighting the roles of government regulators in the Food and Drug Administration and the medium of broadcast television in the 1970s and 1980s. Ilina Singh draws on different sources—namely, advertisements in medical journals and popular magazines from the 1950s to the present—to investigate the marketing of stimulant drugs for children diagnosed with attention disorders. Singh's analysis demonstrates the extent to which stimulant drug advertisements contributed to the medicalization of a seemingly widespread childhood behavior. Marketing campaigns appealed to two different gendered sentiments: the maternal fitness of mothers and the proper behavior of boys. Andrea Tone also notes gendered changes in prescription practices for tranquilizers from the 1950s through the 1970s. More broadly, her chapter examines the dramatic changes that occurred during this period in popular, political, and medical opinions of tranquilizers, and places these changes in the context of the upheavals taking place in American society and culture.

Although the first statin, Mevacor, hit the market in 1987 and Viagra more than a decade later in 1998, the histories of their development and of the medical conceptualization of the conditions they were designed to treat go back several decades. Jeremy Greene explores the history of expert guidelines, commercial clinical trials, and promotional strategies for statins in the expansion of the target populations for cholesterol-lowering programs. He analyzes the roles of the development and marketing of these drugs as part of the ongoing redefinition of the condition known as hypercholesterolemia, elucidating how a pututative risk factor for athrosclerosis has been transformed into a disease. In her study of the erectile dysfunction drug Viagra, Jennifer Fishman also discusses how prescription medicines can spark radically new understandings of and treatments for an age-old complaint. She locates the technological development of Viagra within the historical, social, and scientific contexts of the medicalization of male sexuality. Her analysis helps to explain the blockbuster success of this "lifestyle" drug at the turn of the twenty-first century.

These histories are linked by a number of motifs. Oral contraceptives, hormone replacement therapy, and Viagra are all concerned with sexual and reproductive health. Mood stabilizers, stimulants, and tranquilizers involve psychological and behavioral modification. Prevention is the goal of several of these drug products: oral contraceptives for pregnancy, mood stabilizers for psychiatric episodes, hormone replacement therapy for hot flashes and osteoporosis. While the eight drugs discussed here can be grouped and regrouped according to different characterizations of their targets and purposes, there are three common themes which run through the historical studies in this volume.

The first theme concerns the coconstruction of diseases and treatments. Some of the chapters demonstrate that the definition of pathological states has been closely associated with the development and marketing of new drugs. Fishman argues that the new interpretation of impotence as erectile dysfunction created the social space for Viagra to succeed. Healy describes the creation of both new disease constructs and new drug categories in the case of bipolar disorder and mood stabilizers. Disease categories have also been redefined to accommodate new uses or expanded markets for established drugs. Watkins shows that this was the case for hormone replacement therapy, as its indications expanded from the short-term treatment of hot flashes to the long-term prevention of osteoporosis and cardiovascular disease. Similarly, Singh details the shift in marketing stimulants from a treatment for psychiatric disorders in adults to behavior management and modification in children diagnosed as hyperactive. And Bud comments on the medical and social implications of the wanton prescription and use of antibiotics for nonbacterial diseases.

Drugs have also blurred and shifted the boundaries between normal and abnormal, healthy and pathological. The second theme involves this drug-mediated process of medicalization. Greene links the arrival of the statins to the recalibration of the scale used to determine what level of cholesterol necessitated treatment. Singh argues that stimulant marketing set strict norms for boys' behavior; anything outside those boundaries required medication. The medicalization of what was previously considered part of modern living is exemplified by Tone's demonstration of how the availability of tranquilizers transformed the diagnosis of anxiety into a medical problem in need of pharmaco-therapy. And the medicalization of what was previously considered part of growing old is critical to Watkins's discussion of hormone replacement therapy and menopause and Fishman's story of Viagra and erectile dysfunction. The oral contraceptives studied by Junod also played a role in bringing birth control into the purview of physicians.

The development of drugs used by millions of Americans, potentially for long periods of time, transformed many patients into consumer activists, who demanded to be informed about these drugs, their side-effects, and the conditions they were intended to treat. The third theme addresses the communication of medical information about drugs and diseases. Watkins and Junod focus their attention on the media as health educators about hormone replacement therapy and oral contraceptives, respectively, translating research findings from the technical language of science into the vernacular. The drug companies also used their advertisements as an opportunity to educate the public, starting in the 1980s with print ads and then in 1997 on television commercials; most of the drugs considered in this volume have been heavily promoted by their manufacturers in direct-to-consumer pitches. Finally, given all the publicity generated by pharmaceutical manufacturers, government regulators and commissions, and journalists, drugs became incorporated not simply into medical practice but also into popular culture. This development changed the way Americans thought about health, disease, and modern living, as clearly evinced by Tone's study of the tranquilizers. Bud's concept of rebranding is instructive here, because it captures the shifting fortunes, both anticipated and unanticipated, of specific pharmaceuticals in medical and popular opinion.

Taken together, the chapters in this volume illuminate the ways in which prescription medications have shaped and reshaped modern medicine and life in the past half century. These case studies challenge our conceptions of drugs and therapies, disease and health, treatment and prevention. We hope that the historical perspectives offered here will enrich contemporary conversations about the roles—past, present, and future—of pharmaceuticals in American society and culture.

NOTES

The authors would like to thank James Delbourgo, Howard Kushner, Naomi Rogers, and Nancy Tomes for helpful comments and suggestions in the preparation of this introduction. They also wish to thank the contributing authors as well as Emily Park and Eric Zinner of New York University Press for their support of this volume.

1. Marcia Angell, *The Truth about Drug Companies: How They Deceive Us and What to Do about It* (New York: Random House, 2004), 3; Greg Critser, *Generation RX: How Prescription Drugs Are Altering American Lives, Minds, and Bodies* (New York: Houghton Mifflin, 2005), 2.

2. Londa Schiebinger and Claudia Swan, eds., *Colonial Botany: Science, Commerce, and Politics in the Early Modern World* (Philadelphia: University of Pennsylvania Press, 2004).

3. David Healy, *The Antidepressant Era* (Cambridge, MA: Harvard University Press, 1997), 23. On penicillin and Fleming see Ronald Hare, "New Light on the History of Penicillin," *Medical History* 26 (1982): 1–24 and Gwyn Macfarlane,

Alexander Fleming: The Man and the Myth (New York: Oxford University Press, 2005). On the contingent character of penicillin's development, see Wai Chen, "The Laboratory as Business: Sir Almroth Wright's vaccine program and the construction of penicillin," in Andrew Cunningham and Perry Williams, eds., *The Laboratory Revolution in Medicine* (Cambridge: Cambridge University Press, 1992), 245–92.

4. David S. Jones, "The Health Care Experiments at Many Farms: The Navajo, Tuberculosis, and the Limits of Modern Medicine, 1952–1962," *Bull. Hist. Med.*, 2002, 76: 749–790; on Cold War optimism see James T. Patterson, *Grand Expectations: The United States, 1945–1970* (New York: Oxford University Press, 1996) and Paul Starr, *The Social Transformation of American Medicine: The Rise of a Sovereign Profession and the Making of a Vast Industry* (New York: Basic Books, 1982). See also Allan M. Brandt and Martha Gardner, "The Golden Age of Medicine?" in Roger Cooter and John Pickstone, eds., *Medicine in the Twentieth Century* (Amsterdam: Harwood, 2000), 21–38.

5. John Swann, "FDA and the Practice of Pharmacy: Prescription Drug Regulation before the Durham-Humphrey Amendment of 1951," *Pharmacy in History* 1994, 36 (2): 55–70, and Peter Temin, *Taking Your Medicine: Drug Regulation in the United States* (Cambridge, MA: Harvard University Press, 1980), 51–53.

6. On the evolving relationship between doctors, patients, and prescription drugs see Nancy Tomes, "The Great American Medicine Show Revisited," *Bull. Hist. Med.*, 2005, 79: 627–663.

7. Critser, *Generation RX*, 2; Ray Moynihan and Alan Cassels, *Selling Sickness: How the World's Biggest Pharmaceutical Companies Are Turning Us All Into Patients* (Vancouver and Toronto: Greystone Press, 2005), xiii.

8. Wai Chen, "The Laboratory as Business."

9. Critser, *Generation RX*, 9; Angell, *The Truth about Drug Companies*, 74–93.

10. Harry M. Marks, *The Progress of Experiment: Science and Therapeutic Reform in the United States, 1900–1990* (Cambridge: Cambridge University Press, 1997); Steven Epstein, "Activism, Drug Regulation, and the Politics of Therapeutic Evaluation in the AIDS Era: A Case Study of DDC and the 'Surrogate Markers' Debate," *Social Studies of Science* 27 (1997): 691–726 and Steven Epstein, *Impure Science: AIDS, Activism, and the Politics of Knowledge* (Berkeley, CA: University of California Press, 1996). Also see John Pickstone, ed., *Medical Innovations in Historical Perspective* (Basingstoke: Macmillan, 1992).

11. Marcia Angell, *The Truth about Drug Companies;* David Healy, *Let Them Eat Prozac* (New York: New York University Press, 2002); Moynihan and Cassels, *Selling Sickness.*

12. Andrea Tone, "Introduction" in "Medicine, Technology, and Society," special volume of *History and Technology* 2002, 18 (4): 271–276.

13. Critser, *Generation RX*, 8. On the international dimensions of the industry's growth, see Arthur Daemmrich, *Pharmacopolitics: Drug Regulation in the*

United States and Germany (Chapel Hill: University of North Carolina Press, 2004).

14. See, for instance, David Musto, *The American Disease: Origins of Narcotics Control* (New Haven: Yale University Press, 1973); David Courtwright, *Dark Paradise: A History of Opiate Addiction in America* (Cambridge, MA: Harvard University Press, 1982); John Burnham, *Bad Habits: Drinking, Smoking, Taking Drugs, Gambling, Sexual Misbehavior, and Swearing in American History* (New York: New York University Press, 1993); Caroline Acker and Sarah Tracy, eds., *Altering American Consciousness: The History of Alcohol and Drug Use in the United States, 1800–2000* (Amherst, MA: University of Massachusetts Press, 2004); and Caroline Acker, *Creating the American Junkie: Addiction Research in the Classical Era of Narcotic Control* (Baltimore: Johns Hopkins University Press 2002). Also see Howard Kushner, "Taking Biology Seriously: The Next Task for Historians of Addiction?" *Bull. Hist. Med.*, 2006, 80: 115–145.

15. On the social and political stigmatization of prescription drugs, see Erica Dyck, "Flashback: Psychiatric Experimentation with LSD in Historical Perspective," *Canadian Journal of Psychiatry* 50 (June 2005): 381–88; Andrea Tone, "Listening to the Past: History, Psychiatry, and Anxiety," *Canadian Journal of Psychiatry* 50 (June 2005): 373–80; and Susan Speaker, "From Happiness Pills to National Nightmare," *Journal of the History of Medicine and Allied Sciences* 52 (July 1997): 338–76. Also see Sjaak van der Geest, Susan Reynolds Whyte, and Anita Hardon, "The Anthropology of Pharmaceuticals: A Biographical Approach," *Annual Review of Anthropology* 25 (1996): 153–78.

16. Tone, "Introduction," 273.

17. Roy Porter, "The Patient's View: Doing Medical History from Below," *Theory and Society* 14 (1985): 189. Also see Tomes, "The Great American Medicine Show Revisited," and Nancy Tomes, "Merchants of Health: Medicine and Consumer Culture in the United States, 1900–1940," *Journal of American History* 88 (September 2001): 519–547.

18. Patricia Gossel, "Packaging the Pill," in Robert Bud, Bernard Finn, and Helmuth Trischler, eds., *Manifesting Medicine: Bodies and Machines* (Amsterdam: Harwood, 1999), 105–21; Jones, "The Health Care Experiments at Many Farms"; Lara Marks, "'A Cage of Ovulating Females': The History of the Early Oral Contraceptive Pill Clinical Trials, 1950–1959," in H. Kamminga and S. De Chadarevian, eds., *Molecularizing Biology and Medicine, 1910s-1970s* (Amsterdam: Harwood Academic Publishers, 1998), 221–48.

19. On the history of prescription drug advertising and promotion, see Tomes, "The Great American Medicine Show Revisited." Also see Jeremy A. Greene, "Attention to 'Details': Etiquette and Pharmaceutical Salesmen in Postwar America," *Soc. Stud. Sci.*, 2004, 34: 271–92; Andrea Tone, *Devices and Desires: A History of Contraceptives in America* (New York: Hill and Wang, 2001); Jacalyn Duffin and Alison Li, "Great Moments: Parke, Davis and Company and the Creation of Med-

ical Arts," *Isis,* 1995, 86: 1–29; and Jonathan M. Metzl and Joel D. Howell, "Making History: Lessons from the Great Moments Series of Pharmaceutical Advertisements," *Acad. Med.*, 2004, 79: 1027–32.

20. Moynihan and Cassels, *Selling Sickness*; Ray Moynihan, Iona Heath, and David Henry, "Selling Sickness: the Pharmaceutical Industry and Disease Mongering," *British Medical Journal*, 324 (2002): 886–901; S. Lee and A. Mysyk, "The Medicalization of Compulsive Buying," *Social Studies of Medicine* 58 (2004): 1709–18; Peter Conrad and Valerie Leiter, "Medicalization, Markets and Consumers," *Journal of Health and Social Behavior* 45 (2004): 158–76; Jennifer R. Fishman, "Manufacturing Desire: The Commodification of Female Sexual Dysfunction," *Social Studies of Science* 34/2 (April 2004): 187–218; Jennifer Fosket, "Constructing 'High Risk' Women: The Development and Standardization of a Breast Cancer Risk Assessment Tool," in *Science, Technology, and Human Values* 29 (2004): 291–313; David Healy, "Have Drug Companies Hyped Social Anxiety Disorder to Increase Sales? Yes: Marketing Hinders Discovery of Long-Term Solution," *Western Journal of Medicine* 175 (2001): 364.

Part I

1

ANTIBIOTICS

From Germophobia to the Carefree Life and Back Again
The Lifecycle of the Antibiotic Brand

Robert Bud

The use of "antibiotics" was a defining quality of societies across the world during the second half of the twentieth century. The term itself, in its modern sense of "inhibiting the growth or the metabolic activities of bacteria and other micro-organisms by a chemical substance of microbial origin," was first used in print by Rutgers Professor Selman Waksman in 1942.[1] He had been helping *Biological Abstracts* to identify a general term for an already swiftly multiplying group of compounds just a few months after the first human patients had begun to benefit from refined penicillin. This was the first antibiotic to be widely acclaimed as a therapy, and it would soon raise enormous excitement across the world. Within twenty years, antibiotics passed through three generations of development, with a large number of products, from a variety of chemically different families, and within forty years, the annual global consumption of 17,000 metric tons was sufficient for a treatment for each person on earth.[2]

Antibiotics were iconic products of modern organized research, but they also quickly came to be used in ways determined by vernacular as well as by scientific cultures. Doctors have often used drugs in ways subject more to local custom rather than to "science," and in popular culture antibiotics acquired a strong and coherent meaning, suggesting that they could cure any infection. Amid widening fears of the consequences of overuse, patients and doctors were frequently accused of abusing the drugs, particularly in the later part of the century. At the very end of the century, patterns of use did begin to change in many countries and resort to antibiotics became less automatic in the face of infection. Such distinctive patterns of use present an important challenge to the historiography of antibiotics, which hitherto has been focused more on the triumph of discovery and on the adequacy of regulation.

These twin issues of how antibiotics came to be widely used, and the process of changing practice at the end of the century, will be addressed here by treating antibiotics as brands whose changing social meaning has been constructed jointly by patients, doctors, the pharmaceutical industry, and the press. Although the concept of brand may be more familiar in marketing than in historical analysis, it is appropriate to a product, which grew up in the era of the postwar consumer boom. The concept was itself articulated during the same period of growing consumption in the 1950s and 1960s, in which antibiotics saw their greatest period of development.

Thinking of antibiotics in this way integrates our understanding of their use with the postwar growth of consumer society. It links too with such user-centered conceptions of the social construction of technology as Ruth Schwartz Cowan's "Consumption Junction."[3] A generation of social anthropologists has looked at the "world of goods" and has found in consumption the sources of personal identity. Mary Douglas has suggested that beyond their instrumental value we should consider commodities as tools with which to think.[4] Whereas it is commonplace to observe symbolic personal battles fought out through the choice of cars and clothes, remarkably, therapeutic medicines have not generally been examined in this way.[5] Yet in the case of antibiotics too, we need to combine an understanding of symbolic significance with the appreciation of technical performance.[6]

Although historians of technology have avoided the use of the "brand" in their explanations, sociologists and analysts of culture have not been so self-denying. An early work of French sociologist Jean Baudrillard, *The System of Things,* published in 1968, dealt with the interpretation of consumption. Here Baudrillard focused upon the "brand" as a quality of a product, which condenses the symbolic and instrumental value of a commodity.[7]

Baudrillard also looked beyond the strictly academic literature and cited at length the work of two men who were addressing themselves to business in the interpretation of a new world: the Austrian immigrant to the United States, Ernest Dichter, and the French writer on business, Pierre Martineau. Even Vance Packard, author of the 1957 denunciation of modern marketing entitled *The Hidden Persuaders,* testified to Dichter's effectiveness.[8] Dichter carried out several studies of medicine-taking behavior for different clients during the 1950s, and in these the use of antibiotics loomed large.

Dichter's perceptions were especially acute in his emphasis on the active consumer. Thus, working with a pharmaceutical association he would show a focus group of doctors flashcards of patients talking about antibiotics. He recorded the irritated responses of the professionals to the sight of laypeople deciding for themselves which medicines they needed.[9] In his description of medicine for the California Medical Association in 1950, he suggested that America was going through a revolution as profound as in the reformation in Europe.[10] No longer would lay people defer to experts. He warned his client of the obsolescence of the doctor who could define the terms of engagement with medicines. For Dichter, trained in Vienna in psychoanalysis, users co-constructed brands to help develop their own identities, and he was anxious to warn doctors that like other suppliers they needed to treat clients as partners rather than as supplicants.

As well as the work of Dichter, Baudrillard cited the 1957 volume *Motivation in Advertising. Motives that Make People Buy.* In this book the French business writer Pierre Martineau explored the meaning of instant coffee and the subtle differences in meaning that this drink had acquired from ground coffee. "When people become articulate about coffee," he wrote, "they go way beyond any drab drink which is on the table three times a day like a glass of water." Not only was instant coffee considered "economical," but it was also "suited to young people, rushing to get to work, progressive. This means they are youthful, busy, hardworking, up to date, smart, clever enough to use modern innovations."[11]

Many brands are carefully crafted by suppliers. However, such an active single agency is not necessary. Nor is the concept of brand restricted to particular trademarked varieties. "Swiss watches" or "German cars" have assured qualities which go beyond the marketing of individual marques.[12] Similarly, the category of antibiotics as a whole can function as a "brand" whose assumed qualities of universal efficacy could not be attributed to a particular supplier or even to a single stakeholder such as the medical profession. It can instead be seen, I shall suggest, as growing out of the reaction against moral interpretations of illness early in the twentieth century, the attraction of a technical solution to ill health, and the relationship between doctor and patient in modern societies.

The First Antibiotics

The first antibiotics arrived on the medical scene rapidly in the decade after 1942. The sulfa drugs had only recently been introduced as cures for infection, and these had been dubbed "wonder drugs" by the press— for until then there had been no treatments available to treat most infections. Nonetheless, sulfa drugs were limited in the range of bacteria affected and resistance grew quickly. Now, thanks to antibiotics, it seemed that all bacterial infections could in principle be cured. The accolade of "wonder drug" was attributed by *Time* magazine to penicillin as early as February 1943.[13] To understand the optimism which mounted thereafter and accompanied the antibiotics in their early years, it is worth reflecting on the process of discovery, which yielded such a rapidly widening family of products whose performance developed spectacularly within a few years.

The story of penicillin with its well-known origin in an accident has come to overshadow the subsequent era of organized discovery.[14] Penicillin was of course the first to be used widely, and its numerous well-attested benefits gave enormous encouragement to those who would develop new drugs. The new medicine could cure syphilis and gonorrhea within days and treat wounds, which were already infected and were untreatable with sulfa drugs. It had however particular restrictions on its range and it proved difficult to enhance. Penicillin could not affect the gram-negative bacteria such as salmonella and *E. coli*, which affect the digestive system. Nor was it powerful against the most feared of infectious disease, tuberculosis. Moreover, penicillin had been the product of an accidentally encountered mold and whereas minor modifications could be made, until the late 1950s, it proved difficult to widen substantially its range of activity, either through chemical or biological means. So while production was rapidly enhanced, the fundamental therapeutic benefits of penicillin itself remained, for a decade, difficult to grow.

The family of antibiotics was therefore developed at first through another route. As early as 1939, the drug gramicidin, albeit itself less therapeutically effective than penicillin, had been the successful outcome of a systematic study of antibacterial compounds produced by soil organisms. This was the work of René Dubos of New York's Rockefeller Institute who first announced his success in September 1939. From 1940, a research program in the agricultural faculty of Rutgers University in New Brunswick headed by Waksman who had been Dubos's teacher, sought and found several other antibiotics produced by soil-inhabiting actinomycetes, a family of organisms that lie between single cell bacteria and fungi.[15] This approach would yield many of the diverse products available by the early 1950s.

This research program informed the word play that led Waksman to coin the word "antibiotic." Although he had been prompted by *Biological Abstracts*' taxonomic needs in his selection of the term antibiotic, he had used his opportunity to promote a methodological vision. The sulfa drugs had been developed in the 1930s by routinely screening huge numbers of dyes. As Waksman showed in one of his first papers to use the term, he was emphasizing the potential of identifying therapeutic agents through a new microbiological strategy rather than by simply testing large numbers of haphazardly chosen chemicals.[16] In subsequent years he vehemently resisted others' attempts to modify the meaning of *his* term. As early as 1947, he wrote an article about the history of his coinage and of his defense against those who would have watered it down.[17]

Waksman's category won authority as the difficulty of providing a chemical synthesis of penicillin became plain, and patients, doctors, and microbiologists looked for other equivalent products, and as success attended those expectations. Thus, streptomycin, found in 1943 by a member of Waksman's team, was effective against gram-negative bacteria, and also proved to be the first drug effective against the gram-positive *Mycobacterium tuberculosis* (the cause of tuberculosis). It was perhaps the best known of a long series of important antibiotics identified by the Rutgers team.

The growing expectations that antibiotics would cure an infection, without investigation of its cause, created an opportunity for U.S. pharmaceutical companies seeking new products. As early as 1945, several companies were following Waksman's vision and sifting through dirt and rotting vegetables for new products by the actinomycetes. Success crowned their efforts. A second generation of antibiotics affecting a wide range of both the gram-negative and gram-positive bacteria was known as "broad spectrum," to differentiate them from the more specialized penicillin and streptomycin.

The first of these, Aureomycin (chlor-tetracycline) and chloramphenicol, were introduced in the late 1940s by Lederle and Parke Davis, respectively.[18] As companies piled into making penicillin, even the substantial
corporations, which had been in the front rank of penicillin manufacture,
felt threatened by the competition. They therefore supported extensive
research to find new products and indeed did successfully isolate important new drugs from the late 1940s to the early 1950s.[19] After these successful developments, their value appeared obvious, however, beforehand
it was far from clear that scarce funds be put into speculative screening for
new products at a time when there was pressure on engineers to reduce
costs and improve processes. Nonetheless, the remorseless fall in prices of
existing drugs and the limitations of their performance persuaded even
the larger companies that they had no choice but to develop new products. The experience of this uncertainty and the conflicts between microbiologists on the one hand and engineers on the other were caught by the
author Mildred Savage in her well-researched and evocative novel *In Vivo*,
published a decade later.[20]

The best known of the second generation of antibiotics, tetracycline
(sold under a variety of brand names), was produced a decade after the
launch of penicillin. It had a wide spectrum of action and could be taken
by mouth against many common bacterial assaults. Whereas penicillin
interrupted the process of cell-wall construction in gram-positive bacteria,
tetracycline by contrast disrupts protein and DNA formation within the
bacterium. Although a third generation of antibiotics which emerged in
the 1960s—semisynthetic penicillins, the cephalosporins and the ciprofl-
oxacin family—had a broader range of origins, thereafter the pace of
antibiotic innovation slowed quickly.[21] It was therefore the sudden expansion of the range of antibiotics through soil screening that had sustained
the new brand and whose potential seemed, for a time, to stretch endlessly
into the future.

From a microbiological point of view, the antibiotics found even in
those early years were a diverse family. Though they did of course share an
envelope of medical properties and their effects were limited to the control
of bacteria and excluded viruses, the ways different families interacted
with bacteria, the particular species they affect, and their side-effects were
quite separate. This heterogeneity was important to specialists, but such
microbiological diversity was neglected in the human sphere, by laymen
and even by many doctors, to whom antibiotics proved a remarkably stable medical and cultural category.

The Paradox of Impact

The role of these new drugs in iconic stories about modern health care can be summarized in terms of two contrasting images. Each is a synthesis of many stereotypes, which pervaded the culture of medicine and its history. The first image is very positive: it shows the world changed at a stroke by penicillin and the other antibiotics that followed it. This image was conjured by the widespread adulation of antibiotics in such magazines as *Newsweek* and *Reader's Digest*. Thus, through the *Digest's* pages, the veteran medical writer Paul De Kruif waged his own war during 1943 on those who would keep lifesaving penicillin from American civilians.[22] The term "wonder drug" was used even in U.S. government documents. While other post–World War Two medicines such as cortisone could also be so described, antibiotics were the dominant beneficiaries.[23]

This positive image was characterized by rescue, relief, and reassurance. Images of rescue from complaints otherwise fatal, such as pneumonia, or war wounds were a key part of the landscape of postwar western cultures. They came to be familiar from advertising, often from drug companies but also from newsreels and were summoned up by family stories about how relatives were rescued from otherwise certain death. Relief from pain such as earache was another critical part of the image. Most small children suffer the intense discomfort of *otitis media* and while patients generally recover within a few days without any treatment at all, the relief provided by penicillin meant that 95% of all instances were treated with antibiotics. Sociologists have mapped the anxieties experienced by the parents of sick children and the reassurance a prescription brings.[24]

The other image portrayed antibiotics very much more negatively. Here they were seen as an overrated technical fix whose impact on medicine was exaggerated by doctors overeager to interpret improving health as the result of medicine's increasing power. This image is most closely associated with the British professor of social medicine, Tom McKeown. In a series of articles and books published in the 1970s, he argued that the benefits of modern therapy had been greatly exaggerated and that instead improving nutrition and housing had been responsible for lengthening nineteenth- and twentieth-century lives.[25]

This image took physical form in a graph in McKeown's book, *The Role of Medicine: Dream, Mirage or Nemesis?* Addressing priorities in health research and education, McKeown showed that mortality from pneumonia and tuberculosis followed an exponential decline through the twentieth century. The decline of both illnesses were typically attributed to the antibiotics, but in his graphs the discovery of antibiotics is at most a minor correction and of no apparent long-term significance.[26] Details within this work are flawed: the emphasis on housing and social conditions rather than nineteenth-century public health measures was probably unjustified.[27] The overall message that individual treatment by drugs did not explain long-term decline was however not unique.

McKeown was happy to pay credit to the inspiration of René Dubos, who had turned away from directly antibacterial solutions to disease following the death of his own wife from tuberculosis. Instead, from the early-1950s, Dubos proposed a holistic model of health in which the strength of the patient's system and total well-being were important factors, too. He reported the response to a talk to the American Society for Physicians: half the audience was appalled and half enthused.[28]

Even those disposed to accept the effect of these medicines on life-threatening pneumonia, tuberculosis, and wound infections illness, accepted that they were generally administered for much more minor ills, self-limiting diseases, viruses, and other conditions in which the bacteriological consequences of antibiotics for the course of disease have been very limited. The American military physician, Robert Moser, called such abuse of the drugs "Antibiotic Abandon."[29] There was therefore a paradox between the enthusiastic response of many patients and doctors and experience of a radical change in attitudes to ill-health, on the one hand, and, on the other, the statistics suggesting a general continuation of existing epidemiological trends, and the widespread use of antibiotics where their impact may have been medically marginal.

The Carefree Life

Professional irritation at the "misconstrual" of antibiotics highlights the significance of these drugs as representing a powerful brand, integrating symbolic and instrumental qualities, even though the result appalled Moser and Dubos. This perspective focuses our attention on the relationship between the drug and the interpretations and feelings that illnesses engendered, rather than concentrating on just the physical suffering caused. Above all, the brand came to stand for the technical solution to infection, replacing control through prevention and the long-established moralization of illness. The rich evidence of the past voices represented in the sociological literature on antibiotics, and the suggestive hints in such historical studies on disease as the 1997 symposium *Morality and Health,* suggest that the early branding of antibiotics as wonder drugs acquired special meaning because of the alternative they offered to the moral con-notation of physical suffering.[30]

Although it is perhaps surprising that the sociological studies of the uses of such medicines have been rather separate from the many works on changing attitudes to the morality of disease, the two litera-tures have of course rather different origins. Investigations of patients' attitudes to the immorality of illness had at first been stimulated by interest in class differences in attitudes to medicine, and particularly to chronic complaints, and were then spurred on by the crisis around the moral interpretations of AIDS.[31] In part, therefore, the lack of close ties with the studies of antibiotic use, which have focused on excessive belief in the drugs' efficacy, reflects the distinctions of emphasis between what were "curable" and "incurable" conditions.

The literature does however suggest a close link between the emergence of the new medicines and a release from the moral management of ill-health. The predominant sense that infection was now purely a technical problem was even built into sociological theory shortly after World War Two through Talcott Parsons' 1951 classic of functionalist sociology, *The Social System.* Parsons reflected the shifting attitudes in his view that the patient's responsibility was merely to consult a doctor in the event of ill-ness.[32] Empirical observations reported public perceptions following a similar vein. In a 1958 U.S. survey, conducted by Chicago's National Opin-ion Research Center (NORC), the survey director concluded:

> Ninety per cent of the respondents thought that "doctors today know *a lot more* about treating sicknesses than they did thirty years ago," while an even greater majority thought that "the medicines we have today are *much better* than they were thirty years ago." These views, coupled with the salience to the patient of his physician's technical competence, serve as at least circum-stantial evidence in support of the hypothesis that the esteem in which doc-tors are now held is based to a large degree on pragmatic considerations.[33]

The faith in a technical solution to disease was therefore closely tied to social relations with doctors and a sense of the patient's own empower-ment and freedom from fear. The French sociologists Christiane Herzlich and Janine Pierret concluded in the early 1960s that there was a sense that "real" illness had ceased to be a concern.[34] The British medical sociologist Mildred Blaxter also noted significant changes over time and significant differences in attitudes to health between generations. During the late 1970s, Blaxter studied the attitudes of three generations of women to medicine. The grandmothers born before 1930 were respectful of doctors. They did pay lip-service to modern medicines, but she said this generation rapidly fell back on "mind over matter" models of cure. Their daughters by contrast treated doctors as more-or-less interchangeable tradesmen whose job it was to prescribe medicines.[35] Blaxter has reflected on the attitudes of young mothers of the 1950s:

> They were, however, very conscious that "things had changed." These family stories were usually told in the context of praise for the National Health Service, universally given credit (together with the "medical advances" that had provided a cure for TB and, in their eyes, had been responsible for "conquering" the childhood killers of the past) for major advances in health. G J2 [an interviewee], above, concluded by saying "We hinna got that worry nowadays."[36]

A pioneering study of communication about drugs among doctors conducted in the 1950s by James S. Coleman, Elihu Katz, and Herbert Menzel of Columbia University reported the comment of one doctor: "Nowadays you give a shot of penicillin for pneumonia and cure the patient, but that's no credit to the doctor; all credit goes to the drug. An old doctor wouldn't have had so many patients; he would sit at the patient's bedside until the fever broke."[37] In other words, rather than the practical view of medicines and richly symbolic view of doctors that the medical profession may have preferred, patients were sensitive to the symbolic significance of the medicines and took a strictly practical view of the doctors who controlled the prescriptions to them.

The technical view of illness and pragmatic conception of drugs contrasts with a long-standing moral condemnation of illness which has been explored in general by historians such as Nancy Tomes.[38] A few examples bring the general to life. In a recently published autobiography, the British philosopher of art, Richard Wollheim, provided a Proustian evocation of his middle-class childhood shortly before World War Two.[39] It was a genteel existence with many material luxuries that make his world appear familiar. His was a world of telephones and cars, movie theatres, the ballet, and foreign holidays. There was another presence so strong and distinctive that Wollheim chose it as the title of his book, *Germs*. His mother was sure they pervaded the house and developed an elaborate means of eradicating them. The governess disagreed; she feared them much more outside and protected her charge from their threat in the outdoors. Wollheim reflects on how, for his mother, attitudes to germs and her own personality and identity were so intermingled, he could not imagine her ever changing her mind. Despite their dispute about the location of danger, both mother and governess felt it their moral duty to protect the family against germs.

Similarly, the Israeli writer Amos Oz, remembering his grandmother in the days before World War Two, describes a cultural landscape dominated by the fear of germs and the need for cleanliness. His grandmother expressed her anxiety about a new Middle Eastern environment through obsession with the dangers of infection.[40] These ladies may have been recalled as eccentric, but their attitudes were widespread. Many studies have highlighted the prewar sense that even if illness itself was not morally wrong, failure to have properly prevented it was blameworthy.[41]

From the advantaged position of the years after World War Two and before the advent of AIDS, however, such anxieties came to seem antiquated. We have seen how, with even greater rapidity than their effect on germs, antibiotics had removed the moral curse associated with prewar illness. Infection rapidly became a medical not a moral problem and this was widely credited to antibiotics. The consequence for the culture of illness was to reduce the guilt associated with disease. No longer would syphilis be a mark of Cain on those who had transgressed sexually. The guilt of tuberculosis or pneumonia was no longer to be punished by the penalties imposed on those who had not been looking after their children or themselves. Such illnesses could now be cured, not judged. Data collected in 1989 by Massachusetts sociologists, Margie Lachman and Jackie James, enable the study of attitudes to illness.[42] People who were born from the late 1930s to the early 1950s were considerably less prone to guilt over illness than an older cohort born during the decade prior to 1935. The antibiotic brand was formed amid this revolution.

Agency and the Brand

Who was responsible for the formation of the brand? Through a variety of routes, industry promoted a positive view of their products to doctors and directly to patients. An advertisement for antibiotic drugs around 1950 showed a boy recovering overnight from fever. The child might be relieved of pain, and the parent was reassured he or she would not die. However soon after this advertisement appeared, such direct marketing to patients was banned. Instead companies put huge efforts, amounting to over 20% of their turnover, into marketing to doctors. As companies' dependence on the drugs grew in the 1950s, they threw huge resources at persuading doctors to prescribe their particular antibiotic. The company detailmen dropping in on physicians to encourage them to prescribe trademark drugs became, in some eyes, figures of ill-repute.[43] Even the *Journal of the American Medical Association* was charged with serving as a billboard for expensive advertising.[44]

Marketing by industry was however but a part of the process by which the reputation of antibiotics was made and spread. A classic study of communication explored the epidemiology of the widening grip of one version of tetracycline on physicians' practices across the United States in the 1950s. The Columbia sociologists Coleman, Katz, and Menzel studying the use of one manufacturer's tetracycline by 216 physicians in the Midwest found that within 16 months half were prescribing it more than its predecessors.[45] Within this overall story of rapid acceptance, however, a variety of different patterns of diffusion were described. Those physicians who were well connected with their professional peers showed, in general, a faster rate of acceptance. Within 16 months all the doctors in this subset had tried using the drug. By contrast those doctors who were more oriented toward patient wishes were rather slower to be infected by the tetracycline bug. Only three-quarters of this group of more professionally isolated physicians had used the drug. This suggested that initially, at least, professionally oriented doctors were more aware of the new drug's advantages than doctors sensitive to patients who seemed to be less impressed by the difference between the new drug and its predecessors. In other words, it seemed that peer pressure drove diffusion of new antibiotics.

It is however too simple to "blame" industry or doctors in isolation for the antibiotic brand. A large number of sociological investigations have highlighted the close relationship between attitudes to antibiotic use and relations between doctors and patients. Patients too have often been seen as the key actors in the use of antibiotics. The historian David Adams has surmised that the promise of wartime penicillin symbolized all the hitherto benefits of the "good life" that would follow World War Two.[46] In most countries, physicians have complained that some patients were insistent in their demand for an antibiotic whatever the cause of their complaint.

Certainly, even when drugs were acquired on prescription, patients often maintained a keen interest in what they were taking. Antibiotics proved a special area of negotiation between doctors and patients. Doctors have often felt pressured to prescribe a drug that might well benefit a patient and was unlikely to do any direct harm. As early as 1950, the medical sociologist Elmer Koos reported that one in six of his sample in a small Pennsylvania town had someone in the family who regularly listened to medical programs on the radio, one in five read the daily health column in the newspaper, and one in four read magazine articles about health.[47] Thus, even then patients were also health-care consumers. The availability of the antibiotic brand was a key reference in their consumption of such care.

Some physicians have reported that they have prescribed antibiotics not so much because they anticipated an effect on bacteria but because they wanted to build a relationship with the patient. In Britain, the medical sociologist Nicky Britten investigating patient behavior in the 1990s has shown that a considerable proportion of patients have entered doctors' practices not wanting but expecting an antibiotic prescription. Meanwhile, the hurried physician with inadequate time to ascertain the patient's actual wishes has decided to prescribe an antibiotic in the expectation that the public would want one.[48]

It was not only in wealthy countries that antibiotics developed a high reputation. In Kenya, the respected Vienna-born doctor J. P. Loefler, in a prize winning article published by *The Lancet* blamed the pattern of antibiotic consumption that he was witnessing on "the unholiest of alliances": consumer demand based on misinformation, mercantile interest, and the insecurity and cynicism of the middlemen.[49] Other reports from Africa have reported antibiotics used as cure-alls for all conditions.[50]

Again, in developing countries as in the United States, the product affected the mind as well as the body and bacteria. Such meanings came not just from enthusiastic salesmen: the attraction of injected varieties in some countries expressed the interaction of local cultures with the injunctions of medicine. The very act of prescribing could have a therapeutic effect. A study in the 1990s found that in such developing countries as Indonesia, Pakistan, Ghana, and Uzbekistan 60% of all consultations with a doctor ended in an injection, often of antibiotics.[51]

Despite the discomfort, administration of antibiotics by injection was particularly popular in many countries in both Africa and Asia. Since the 1920s when mass treatment of yaws with bismuth and arsenic compounds had begun, the process of injecting had been associated with the potency of western medicine (despite the occasionally fatal consequences of these particular treatments). Moreover, the individuality of the injecting process has been linked to the growing attraction of individualist culture in countries such as Uganda.[52] Like other brands, using an antibiotic helped patients define themselves within society as a whole.

The image of the antibiotic brand faced several major challenges in its first thirty years. Rather than devaluing the brand, however, these problems were diverted into skepticism of, or even hostility to, the drug companies that made the drugs and profited from them. Cost was a repeatedly raised issue. New products began as expensive luxuries even if some quickly became cheap commodities. Penicillin was the model of a scarce and valued product whose price collapsed. Similarly streptomycin emerged in the early postwar years, as a product so expensive that the British were only willing to import it after the most rigorous clinical trial ever conducted to that time.[53] Its price too fell rapidly. On the other hand, the price of the new broad-spectrum antibiotics after an initial fall was maintained at a price widely considered to be "high" and an enduring issue even for American consumers in the 1950s and 1960s.

The cost of antibiotics was challenged by U.S. senators from the early 1950s and they pushed the Federal Trade Commission to investigate the industry and its profits.[54] This assault was followed by an investigation of four classes of drugs (antibiotics, corticosteroids, tranquilizers, and oral antidiabetics) conducted by the Congressional Subcommittee on Antitrust and Monopoly under Senator Kefauver.[55] This committee held a series of hearings on antibiotics from 1959 to 1960, inspired by stories of unaffordable drugs and evidence of excess profits. The federal government would have rejected Kefauver's subsequent bill regulating the industry, had the terrifying tale of "thalidomide" not emerged in mid-1962. Although this sedative (not an antibiotic) was never released for sale in the United States, the press revealed how close the public had been to acquiring what was clearly a dangerous drug affecting both fetuses and adults.[56] The Food and Drug Administration was rapidly given considerable new powers to regulate the introduction of new drugs in the United States, but the reputation of antibiotics themselves had not been affected for most people.

The antibiotic brand itself was not affected even by fears about safety. Penicillin's lack of toxicity for most patients had been a dramatic feature of its early use. Nonetheless, in subsequent years, allergic reactions ranging from dermatitis to sudden death were encountered.[57] Increased purification addressed some of these, but there remained a small proportion of patients allergic to penicillin. Worse, the early broad-spectrum drug chloramphenicol widely used for minor infections, and the major early treatment for typhoid, was shown to lead occasionally to a variety of leukemia.[58] The dangers of chloramphenicol were also exposed at the hearings of Senator Kefauver's Committee. Again, the problems exposed were blamed not on the drugs but on the companies that promoted them without providing adequate warnings.

Notwithstanding these challenges, therefore, the antibiotic brand gave meaning to the general historical experience of improving health, and interpretations of the historical experience became themselves part of the brand. The emergence of antibiotic resistance, however, made this reassuring condensation of instrumental and symbolic values unsustainable. In the 1990s, there was a widening attempt to transform what had become one of the strongest brands in the world and to persuade the public that, notwithstanding earlier beliefs, these were not miracle drugs.

Resistance

From the 1940s, public health officials worried about the rise of resistance to penicillin. Resistant staphylococci became a substantial problem in hospitals worldwide in the 1950s.[59] At first this seemed to be a problem restricted to a single family of drugs. However, by the early 1960s, Japanese studies of stretches of DNA transferred between bacteria (at first called episomes then plasmids) tended to suggest both that resistance could move between bacteria and that organisms could demonstrate resistance to quite different families of antibiotic.[60] In other words, the problem of resistance could be seen in terms of the entire class of antibiotics and not just as the entirely separate difficulties of individual classes of drug. During the 1960s, multiply-resistant gram-negative bacteria such as *E. coli* and salmonella became a major concern. In the late 1970s and 1980s, resistant streptococci including the common pneumonia-causing agents pneumococci became a growing worry.[61]

At first, the pharmaceutical companies responded by trying to develop new agents. Methicillin, an early semisynthetic penicillin, was launched in 1960 specifically to counter staphylococci resistant to earlier antibiotics. Staphylococci resistant to methicillin, MRSA (Methicillin Resistant *Staphylococcus aureus*), were, however, selected from naturally occurring strains. They were first encountered in the 1960s and they were being widely discussed by infection specialists by the 1980s, though for unknown reasons their incidence fell for a time until rising epidemically in the 1990s. Gram-negative bacteria also showed immunity to antibiotics and again a partial technical solution was developed. Amoxicillin, the broad spectrum semisynthetic penicillin, was blended with a compound that lapped up the betalactamase enzyme produced by resistant organisms to create the drug known as Augmentin.[62]

In the face of the growing challenges, rewards from traditional approaches to antibiotic development were diminishing. From the early 1960s until the late 1990s, not a single fundamentally new family was introduced. The century ended, however, with an optimism renewed on the back of new discovery technology: the introduction of gene sequencing, genetic engineering, and computerized chemical techniques. A bacterial genome would be sequenced, a point at which reproduction could be interrupted, deduced, and a variety of likely candidate poisons tested. The SmithKline Beecham Company announced that it was pursuing a "Manhattan-micro" project to find new antibiotic bombs that would eradicate the enemy.[63] Nonetheless, with little substantial success and apparently small financial incentives some research programs were run down early in the twenty-first century. The prospect of a technical solution to the problem of resistance seemed to recede, and with it the sustainability of the antibiotic brand came into question.

Rebranding Antibiotics

The threats of antibiotic resistance in the bacterial population worsened considerably from the late 1980s. MRSA was becoming a steadily more challenging problem in hospitals in the United States and the United Kingdom. Moreover resistant strains of the pneumonia-causing streptococcus, which had been slow to emerge, were now becoming common in France and Spain.[64] Challenging as these threats were, they were given much greater prominence because public health campaigners linked them to the new and terrifying category of "emergent infectious diseases." This phrase first became well known by means of a 1992 Institute of Medicine Conference in Washington.[65] It became a way of reimagining bacterial infection from the postwar nuisance back to a worrying, if manageable threat. A host of initiatives during the 1990s would reinforce the link between the scourges of terrifying prions (which caused Creutzfeld-Jacob Disease (CJD)), viruses that caused potentially species-threatening diseases, and such new threats as MRSA and resistant pneumococci and *E. coli*. It must be emphasized that the grouping of such bacterial threats with viruses that were biologically completely separate was intended to raise alarm about the consequences of the misuse of antibiotics and was not the necessary consequence of biological taxonomy. By lumping MRSA infection together with the already feared AIDS and the terrifying CJD, much greater attention was focused on the threat posed by antibiotic-resistant bacteria.

The context had been set already by a gradually growing remoralization of disease in the 1970s. Such chronic problems as heart disease, stroke, and lung cancer were of increasing importance now that acute infection could often be controlled. These, however, could often not be resolved through drugs and instead improved personal behavior was seen as the solution. Eating properly, giving up smoking, and exercise were the responsibility of the individual.[66]

The emergence of AIDS early in the 1980s gave further impetus toward a moral interpretation of disease. At first principally associated with promiscuous homosexuality the general public was happy to cast blame. The essayist Susan Sontag summarized the mood of the time. "Medicine changed mores. Illness is changing them back," she commented in her book *AIDS and Its Metaphors*.[67] HIV proved to be just one of an apparently new host of threatening organisms. Marburg virus, Lassa fever, Hepatitis C, "flesh-eating virus," and CJD entered the public imagination in the late 1980s.

Historian Nancy Tomes has pointed out that the "gospel of germs was born again in the age of AIDS."[68] The new disease generated an underlying fear, which scientists could try to shape to affect widespread cultural change. The evidence of the Boston study suggested that respondents born after 1951 were somewhat more likely than their parents to accept blame for illness.[69]

New attitudes were promoted through a rare alliance between the press and science. When antibiotics were being developed, such leaders of the research community as Howard Florey had avoided the press and even Alexander Fleming, who allowed himself to be widely photographed and quoted, had not actively used journalists to promote particular viewpoints. By contrast in the 1990s, specialist scientists worked willingly and supportively with journalists and writers who promoted their message through popular works. Thus, the help of the Nobel-Prize winning president of Rockefeller University, Joshua Lederberg, was widely acknowledged. One of his former students, New York Times journalist Laurie Garrett, researched and wrote a substantial if accessible work published in 1994 as the Coming Plague.[70] A more popular warning of the imminent threat of new diseases was the Hot Zone widely available in airports and bookstands.[71] Capitalizing on the success of this book, Warner Brothers made a successful movie Outbreak. Viruses even pervaded computers. The news of global challenges resonated with the particular concerns in individual countries. In Britain, for instance, the early 1990s were marked by the anxiety over variant CJD and a short-lived panic over the threat of "flesh-eating virus."[72]

The years from 1995 to 1997 saw a great wave of public warnings supported by scientists. Under the umbrella of emergent infectious diseases, the threat of multiplying viruses merged with antibiotic-resistant bacteria. As infectious disease was reevaluated by the public, antibiotics were also effectively rebranded. By 1997, politicians could feel the mood changing. In a variety of countries, parliaments and governments sponsored inquiries and meetings. Britain's House of Lords held one of the most high profile of detailed inquisitions into antibiotic use, threatening that the alternative to reducing antibiotic use was to "return to the bad old days of incurable diseases before antibiotics were available."[73]

In September 1998, the European Commission sponsored a major meeting in Copenhagen on antibiotics and the emergence of resistant organisms, leading to more systematic surveillance and reporting of the use of the drugs across Europe.[74] The following May, The World Health Organization sanctioned a report on the control of global antibiotic use. The 2001 report put education of the public into how and when to take antibiotics at the head of its list of recommendations.[75]

From 1997 national organizations began to campaign actively to change public perceptions. In Britain, advertisements used in a 1999 campaign reflected the 1940s aspirations by using 1940s style graphics but mutated them. The modern posters announced what antibiotics would not do.[76] A new initiative created by the private charity "Doctor Patient Partnership" produced a brochure, "not a miracle cure."[77] Instead of giving a patient a prescription or alternatively making an unhappy person leave without anything at all, the doctor could hand over this leaflet.[78] This campaign appeared to show success: the use of antibiotics by family practitioners fell in Britain by 25% between 1995 and 2000, and there was evidence of similar falls elsewhere.[79] In California a statewide initiative led to the "Alliance Working for Antibiotic Resistance Education" (AWARE) which sought to collaborate with the public to reduce abuse.

This activity did not mean that patients necessarily understood entirely the messages they were given; indeed, there was continuing confusion between the resistance of bacteria and the resistance of patients. A follow-up on a 2003 British government campaign found that "there was an increase to 82% of people who agreed with the statement that taking too many antibiotics weakens the body's resistance."[80] While their microbiology had confused the resistance of the patient and of bacteria, the result could be seen as a radical rebranding of the antibiotic. No longer the first resort in the case of infection and the alternative to relying on a body's defenses, it was promoted as the last resort to be used only in the case of a failure of self-defense.

Patients in several European countries became used to not receiving a prescription for an antibiotic in response to conditions such as "acute red ear," a middle ear infection. On other occasions, prescriptions have been offered but parents of sick children are asked to wait three days before collecting them. In one British study, three-quarters of the parents were happy with the wait-and-see approach and were also more willing to wait before seeing a doctor on a future occasion. In a later U.S. study, half the families would be inclined to withhold antibiotics in the future.[81] Some suffering was taken as "acceptable," and the changing views of antibiotics implied limitations of the doctor as an agent of relief.

The history both of the introduction of antibiotics and of their frustration through resistant bacteria such as MRSA has often been recounted as direct engagements with biochemicals rather than as human dramas mediated by expectations of, and experiences with, a brand. While the brand cannot exhaust the meaning and potential of antibiotics, it provides a powerful and hitherto overlooked means of telling the history of such drugs and possibly of other technologies.

Here the concept of brand has served as a tool for understanding the transformation in the culture of both antibiotics and infections since World War Two. As this chapter has shown, the image of antibiotics as wonder drugs had emerged as part of a complex of new developments in relation to the medicalization of disease and the changing relationships with doctors. Infection was turned into a technical rather than a moral problem by means of antibiotics. As resistance to antibiotics grew in bacterial populations, public health promoters were faced by the challenge of changing the public's views on antibiotics. Their numerous campaigns can be seen as an exercise in rebranding. This would require changes in attitudes to a particular drug—but also toward ill health and the role of medicine.

NOTES

I am grateful to Professor Douglas Eveleigh for his comments on an earlier draft. The research for this article was supported by the Science Museum.

1. Selman Waksman, "What is an Antibiotic or Antibiotic Substance?" *Mycologia* 39 no. 5 (1947): 565–69 lays out the early history of the word from Waksman's perspective. The two articles he and his team published that year promoting the concept of "antibiotic" were S. A. Waksman and H. M. Woodruf, "Selective Antibacterial Action of Various Substances of Microbial Origin," *Journal of Bacteriology* 44(1942): 373–84; S. A. Waksman, E. Horning, M. Welsch, and H. B. Woodruf, "Distribution of Antagonistic Actinomycetes in Nature," *Soil Science* 54(1942): 281–96.

2. R. H. Liss and F. R. Batchelor, "Economic Evaluations of Antibiotic Use and Resistance—a Perspective Report of Task Force 6," *Reviews of Infectious Diseases* 9 (May-June 1987): S297–S312.

3. The classic statement of SCOT is Wiebe A. Bijker, Thomas P. Hughes, and Trevor Pinch, eds., *The Social Construction of Technological Systems: New Directions in the Sociology and History of Technology* (Cambridge, Mass: MIT Press, 1987). See also N. E. J. Oudshoorn and T. J. Pinch, eds., *How Users Matter: The Co-construction of Users and Technology* (Cambridge, Mass: MIT Press, 2003). See too Sheila Jasanoff, ed., *States of Knowledge: The Co-production of Science and Social Order* (New York: Routledge, 2004).

4. See also Mary Douglas and Baron Isherwood, *The World of Goods: Towards an Anthropology of Consumption* (New York: Basic Books, 1979).

5. The literature on this era is, of course, enormous. Recent contributions include Martin Daunton and Matthew Hilton, *The Politics of Consumption: Material Culture and Citizenship in Europe and America* (Oxford: Berg, 2001); Lizabeth Cohen, *A Consumers Republic: The Politics of Mass Consumption in Postwar America* (New York: Knopf, 2003). Neither of these works, however, deal with the consumption of medicinal drugs.

6. The independence of a brand from a manufacturer's intent can be seen most easily in the case of automobiles. The brand consultant and writer Wally Olins points out that Volkswagen was initially "baffled and discomfited" by the North American interpretation of their fine automobiles as lovable "beetles." Wally Olins,

On B®and (London: Thames and Hudson, 2003). I am very grateful to Mr. Olins for the time he spent discussing the concept of brand with me. Also see as an introduction to a vast literature, Giep Franzen and Margot Brouwman, *The Mental World of Brands—Mind, Memory and Brand Success* (Henley on Thames, Oxon: World Advertising Research Centre, 2001). One distinguished exception to the general rule that historians have overlooked the "idea" of antibiotics is James C. Whorton, "'Antibiotic Abandon': The Resurgence of Therapeutic Rationalism," in *The History of Antibiotics: A Symposium,* edited by John Parascandola (Madison, Wisc.: American Institute for the History of Pharmacy, 1980), 125–36.

7. Jean Baudrillard, *Le système des objets* (Paris: Gallimard, 1968).

8. Vance Packard, *The Hidden Persuaders* (Harmondsworth, Eng.: Penguin, 1960; first published 1957), 32–34.

9. Institute for Motivational Analysis, *A Research Study on Pharmaceutical Advertising* (New York: Pharmaceutical Advertising Club, 1955), 25.

10. Institute for Motivational Analysis, *A Psychological Study of the Doctor-Patient Relationship,* submitted to California Medical Association, May 1950, p. 6.

11. Pierre Martineau, *Motivation in Advertising: Motives that Make People Buy* (New York: McGraw Hill, 1957), 54.

12. Thus, "France" has been consciously developed as a brand of travel destination. Olins, *On B®and.*

13. "Penicillin," *Time* (8 February 1943): 41.

14. Recent accounts have been published by Eric Lax, *The Mould in Dr Florey's Coat* (New York: Henry Holt, 2004); Kevin Brown, *Penicillin Man: Alexander Fleming and the Antibiotic Revolution* (Stroud, Glos: Sutton, 2004). See also Robert Bud, *Penicillin: Triumph and Tragedy* (Oxford: Oxford University Press, 2007).

15. Hubert Lechevalier, "The Search for Antibiotics at Rutgers University" in *The History of Antibiotics,* edited by Parascandola, 113–24.

16. S. A. Waksman and E. S. Horning, "Distribution of Antagonistic Fungi in Nature and Their Antibiotic Action," *Mycologia* 35 no. 1(1943): 47–65.

17. Waksman, "What is an Antibiotic or Antibiotic Substance?"

18. Thomas Maeder, *Adverse Reactions* (New York: William Morrow, 1994); B. M. Duggar, "A Product of the Continuing Search for New Antibiotics," *Annals of the New York Academy of Sciences* 51 (1948–51): 177–81; Henry Welch and Felix Marti-Ibanez, *The Antibiotic Saga* (New York: Medical Encyclopedia, 1960).

19. Walter Sneader, *Drug Discovery: A History* (Chichester, West Sussex: John Wiley, 2005), 300–13.

20. Mildred Savage, *In Vivo* (London: Longmans, 1965; first published 1964). I am grateful to Ms. Savage for the chance to discuss her work.

21. E. M. Tansey and L. M. Reynolds , eds., *Post-penicillin Antibiotics: From Acceptance to Resistance?* Wellcome Witnesses to Twentieth Century Medicine 6 (London: Wellcome Institute for the History of Medicine, 2000).

22. Paul de Kruif, *A Sweeping Wind* (New York: Harcourt, 1962), 208.

23. See, for example, the use in such a legal context as "In the matter of American Cynamide, Bristol Myers, Bristol Laboratories, Charles Pfizer, Olin Mathieson, Upjohn Co," para 4, U.S. FTC folder No 7211.1.1, Box 199, RG 122, U.S. National Archives.

24. Joe Kai, "What Worries Parents When Their Preschool Children are Acutely Ill and Why: a Qualitative Study," *BMJ* 313 (19 October 1996): 983–86.

25. See, for instance, T. McKeown, *The Role of Medicine: Dream, Mirage or Nemesis?* (London: Nuffield Provincial Hospitals Trust, 1976).

26. Ibid. 92–101.

27. J. Colgrove, "The McKeown Thesis: A Historical Controversy and Its Enduring Influence," *American Journal of Public Health* 92(May 2002): 725–29.

28. René Dubos, "The Philosopher's Search for Health," *Transactions of the Association of American Physicians* 66 (1953): 31–41.

29. R. H. Moser, *Diseases of Medical Progress* (Springfield, Ill.: Charles Thomas, 1959), 8.

30. Paul Roizin and Allan Brandt, eds., *Morality and Health* (London: Routledge, 1997).

31. See, for example, R. Pill and N. C. H. Stott, "Concepts of Illness Causation and Responsibility: Some Preliminary Data from a Sample of Working-Class Mothers," *Social Science and Medicine* 16 no. 1 (1982): 43–52; Alan Radley, *Making Sense of Illness* (London: Sage 1994); Arien Mack, ed., *In Time of Plague: The History and Social Consequences of Lethal Epidemic Disease* (New York: New York University Press, 1991). See also Allan Brandt, "Behavior, Disease and Health in the Twentieth Century United States: the Moral Valence of Individual Risk," pp 53–77, in *Morality and Health*, edited by Roizin and Brandt.

32. Talcott Parsons, *The Social System* (London: Routledge and Kegan Paul, 1951).

33. Jack Feldman, "What Americans Think about Their Medical Care" in American Statistical Association, *Proceedings of the Social Statistics Section 1958* (Washington, D.C.: American Statistical Association, 1959): 102–05.

34. Claudine Herzlich and Janine Pierret, *Illness and Self in Society,* translated by Elborg Forster (Baltimore: Johns Hopkins University Press, 1987), 197–99.

35. Mildred Blaxter, *Mothers and Daughters: A Three Generation Study of Health Attitudes and Behaviour* (London: Heinemann, 1982).

36. Mildred Blaxter, "Why Do Victims Blame Themselves" in *Worlds of Illness: Biographical and Cultural Perspectives on the Worlds of Illness,* edited by Alan Radley (London: Routledge, 1993), 124–42, p. 129.

37. J. S. Coleman, E. Katz, and H. Menzel, *Medical Innovation: A Diffusion Study* (Indianapolis: Bobbs-Merrill, 1966), 12.

38. Nancy Tomes, *The Gospel of Germs: Men, Women, and the Microbe in American Life* (Cambridge, Mass.: Harvard University Press, 1998).

39. Richard Wollheim, *Germs: A Memoir of Childhood* (London: Waywiser Press, 2004).

40. Amos Oz, *A Tale of Love and Darkness* (London: Harcourt, 2004).

41. See, for example, Mildred Blaxter, "Whose Fault Is It? People's Own Conceptions of the Reasons for Health Inequalities," *Social Science and Medicine* 44 no. 6 (1997): 747–56; Pill and Stott, "Concepts of Illness Causation and Responsibility," idem, "Preventive Procedures and Practices among Working-Class Women: New Data and Fresh Insights," *Social Science and Medicine* 21 no. 9 (1985): 975–93; Margot Jefferys, J. H. Brotherton, and Ann Cartwright, "Consumption of Medicines on a Working-Class Housing Estate," *British Journal of Preventive and Social Medicine* 14 (April 1960): 64–76.

42. This research used the Health and Personal Styles, 1989 data set [made accessible in 1997, original paper records and electronic data file]. These data were collected and donated by Dr. Margie Lachman and Dr. Jackie James and are made available through the archive of the Henry A. Murray Research Center of the Radcliffe Institute for Advanced Study, Harvard University, Cambridge, Massachusetts [Producer and Distributor]. Question MHLC 8. Whereas on a scale of 1–6 (6 highest agreement, and 1 highest disagreement), the responses of 8 of the 17 born 1924–1931 went from "slight" disagreement to strong agreement (3–6), only one quarter of the 24 born 1932–1949 fell into this range.

43. Jeremy Greene, "Attention to 'Details': Etiquette and the Pharmaceutical Salesman in Postwar America," *Social Studies of Science* 34 no. 2 (2004): 271–92.

44. Estes Kefauver, *In a Few Hands: Monopoly Power in America* (New York: Pantheon Books, 1965), 71–77.

45. Coleman, Katz, and Menzel, *Medical Innovation.* Although they do not specify the drug whose dissemination they studied, Temin suggested that it was a Lederle version of tetracycline, Peter Temin, "Physician Prescribing Behavior: Is there Learning by Doing?" in *Drugs and Health: Economic Issues and Policy Objectives,* edited by Robert B. Helms (Washington, D.C.: American Enterprise Institute for Public Policy Research, 1981), 173–82. The several publications that came out of the Coleman, Katz, and Menzel research are summarized by Everett M. Rogers, *Diffusion of Innovations,* fifth edition (New York: Free Press, 2003), 65–68.

46. David P. Adams, *The Greatest Good to the Greatest Number: Penicillin Rationing on the American Home Front, 1940–1945* (New York: Peter Lang, 1991), 134–42.

47. E. L. Koos, *The Health of Regionsville* (New York: Columbia University Press, 1954), 116.

48. Nicola Britten, "Lay Views of Medicines and Their Influence on Prescribing: A Study in General Practice," PhD diss., London University, 1996.

49. J. P. Loefler, "Microbes, Chemotherapy, Evolution, and Folly," *Lancet* 348 (21 December 1996): 1703–04.

50. Dianna Melrose, *Bitter Pills: Medicine and the Third World Poor* (Oxfam: Public Affairs Unit, 1983).

51. Melinda Pavin, Tafgat Nurgozhin, Grace Hafner, Farruh Yusufy, and Richard Laing, "Prescribing Practices of Rural Primary Health Care Physicians in Uzbekistan," *Tropical Medicine and International Health* 8 (February 2003): 182–90.

52. S. R. Whyte, "Penicillin, Battery Acid and Sacrifice: Cures and Causes in Nyole Medicine," *Social Science and Medicine* 16 no. 23 (1982): 2055–64.

53. Alan Yoshioka, "Streptomycin, 1946. British Central Administration of Supplies of a New Drug of American Origin with Special Reference to Clinical Trials in Tuberculosis," PhD diss., University of London, 1998.

54. Federal Trade Commission, *Economic Report on Antibiotics* (Washington, D.C.: U.S. Government Printing Office, 1957).

55. Robert Bud, "Antibiotics, Big Business, and Consumers: The Context of Government Investigations into the Postwar American Drug Industry," *Technology and Culture* 46 no. 2 (2005): 329–49.

56. Richard E. MacFadyen, "Thalidomide in America: A Brush with Tragedy," *Clio Medica* 11 no. 2 (1976): 79–93.

57. Gordon T. Stewart and John P. McGovern , eds., *Penicillin Allergy: Clinical and Immunologic Aspects* (Springfield, Ill.: Charles C. Thomas, 1970).

58. Thomas Maeder, *Adverse Reactions* (New York: William Morrow, 1994).

59. Craig H. Steffen, "Penicillins and Staphylococci: A Historical Interaction," *Perspectives in Biology and Medicine* 35 no. 4 (1991–1992): 596–608; Donald I. McGraw, "The Golden Staph: Medicine's Response to the Challenge of the Resistant Staphylococci in the Mid-Twentieth Century," *Dynamis* 4 (1984): 219–37; R. E. O. Williams, "Investigations of Hospital-acquired Staphylococcal Disease and its Control in Great Britain," *Proceedings of the National Conference on Hospital-acquired Staphylococcal disease, sponsored by US PHS and NAS, Atlanta Georgia* (Washington, D.C.: US DHEW, CDC, 1958), 11–29.

60. T. Watanabe, "Infectious Drug Resistance," *Scientific American* 217 no. 6 (1967): 19–28.

61. Ricki Lewis, "The Rise of Antibiotic-Resistant Infections," *FDA Consumer Magazine* (September 1995) http://www.fda.gov/fdac/features/795_antibio.html accessed 14 November 2005.

62. G. N. Rolinson, "Evolution of β-Lactamase Inhibitors," *Review of Infectious Diseases* 13 (July-August 1991): Supp: 727–32.

63. "Memorandum by SmithKline Beecham Pharmaceuticals," *Resistance to Antibiotics and Other Antimicrobial Agents, Seventh Report of the Select Committee on Science and Technology (Lords). Evidence.* HL Paper 81-II, Session 1997–98 pp. 472–85.

64. J. Casal, A. Fenol, M. D. Vicioso, and R. Munoz, "Increase in Resistance to Penicillin in Pneumococci in Spain," *Lancet* 1 (1 April 1989): 735; D. Felmingham

and J. Washington, "Trends in the Antimicrobial Susceptibility of Bacterial Respiratory Tract Pathogens—Findings of the Alexander Project 1992–1996," *Journal of Chemotherapy* 11 (February 1999): Suppl 1:5–21; F. Baquero, "Pneumococcal Resistance to Beta-lactam Antibiotics: A Global Geographic Overview," *Microbial Drug Resistance* 1 (Summer 1995): 115–20.

65. J. Lederberg, R. E. Shope, and S. C. Oaks , eds., *Emerging Infections: Microbial Threats to Health in the United States* (Washington, D.C.: National Academy Press, 1992).

66. Paul Roizin, "Moralization," in *Morality and Health*, 379–40.

67. Susan Sontag, *AIDS and Its Metaphors* (New York: Farrar Straus Giroux, 1989), 76.

68. Tomes, *The Gospel of Germs,* 255.

69. Health and Personal Styles, 1989 data set.

70. Laurie Garrett, *The Coming Plague: Newly Emerging Diseases in a World Out of Balance* (New York: Farrar Straus, 1994).

71. R. Preston, *The Hot Zone* (New York: Random House, 1994).

72. Bernard Dixon, "Killer Bug Ate My Face," *Current Biology* 6 (1 May 1996): 493; Tim Radford, "Medicine and the Media: Influence and Power of the Media," *Lancet* 347 (1 June 1996): 1533–35.

73. Lord Soulsby, 16 November 1998, *Parliamentary Debates (Lords)* 5th series, 594 (1997–98), col. 1044.

74. European Union Conference on "The Microbial Threat," 9–10 September 1998, Copenhagen, Denmark.

75. WHO, *Global Strategy for Containment of Antibiotic Resistance* (Geneva: WHO, 2001).

76. The posters by the agency Ogilvy and Mather featured a character called "andybiotic."

77. Doctor Patient Partnership, "Antibiotics Not a Miracle Cure!" See *British Medical Journal* 315 (8 November 1997): 1240.

78. I am grateful to Kristin McCarthy, chief executive of DPP for explaining the background to the DPP leaflet.

79. Azeem Majeed, "Reducing Antibiotic Prescriptions," *Canadian Medical Association Journal* 167(15 October 2002): 850.

80. IPSOS UK Research. I am grateful to Ogilvy and Mather for permission to see and quote this document.

81. Paul Little, Clare Gould, Ian Williamson, Michael Moore, Greg Warner, and Joan Dunleavey, "Pragmatic Randomised Trial of Two Prescribing Strategies for Childhood Acute Otitis Media," *BMJ* 322(10 February 2001): 336–42; David M. Spiro, Khoon-Yen Tay, M. Douglas Baker, Donald H. Arnold, and Eugene D. Shapiro, "Wait-and-See Antibiotic Prescription for the Treatment of Acute Otitis Media: A Randomized, Controlled Trial," *Academic Emergency Medicine* 12 (2005): Suppl 1: 19.

2

MOOD STABILIZERS

Folie to Folly

The Modern Mania for Bipolar Disorders and Mood Stabilizers

David Healy

The Foreground

The first use of the term mood stabilizer in the title of a psychiatric article thrown up by Medline or other literature databases is from 1985. In that year an article by Guy Chouinard and colleagues from Montreal commented on the use of estrogen and progesterone as a mood stabilizer.[1] The word is mentioned again in a handful of article titles through 1994, when there is an exponential explosion in its use. By 2001, over a hundred articles a year feature mood stabilizer in their titles. Everybody was talking about mood-stabilizing drugs, even though reviews make it clear that the academic community had still not worked out what the term means.[2] Everybody was acting as though everybody knew what a mood stabilizer was and the assumption was that patients with bipolar disorders need these drugs and perhaps should only be given these drugs rather than any other psychotropic drugs. Mood stabilization had become the trump suit in the psychotropic pack.

As this swarm of publications on mood stabilizers descended, estimates of the prevalence of manic-depressive disease, or folie circulaire as it had been christened by Jean-Pierre Falret in 1851, expanded in parallel, rising from prevalence estimates of 0.1% of the population to figures of 5% or more.[3] An alternate name for the illness had been coined in the mid-1960s and this new name, bipolar disorder, came into vogue in the 1990s, along with a Bipolar Journal, and a slew of bipolar societies and annual conferences. Along with increased estimates for frequency of bipolar disorders, the threshold for the perceived age of onset of the illness dropped so that children aged one or two years old were increasingly diagnosed as having bipolar disorder and were increasingly likely to be treated with mood stabilizers, some of which are the most toxic drugs available to psychiatry. As of the time of this writing, this phenomenon is happening almost exclusively in North America.[4]

The erection of this bipolar edifice qualifies for categorization as a folly, whereby a folly is meant something that was or could readily have been recognized at the time as a mistake or an absurdity.

The Background: Lithium

Mood stabilization in the twenty-first century is about what used to be termed prophylaxis. The earliest ideas that an agent might be in some sense prophylactic against further episodes of a psychiatric disorder were linked with the use of lithium in the nineteenth century. First discovered in 1817, lithium was subsequently found in spa waters and linked to their health-maintaining properties. When lithium was found to dissolve urates in urine, it began to be used in rheumatism, gout, and related conditions, as gout was caused by uric acid deposits in the joints. Individuals predisposed to rheumatoid and related conditions, it was hypothesized, had a predisposition to produce excess uric acid. One of the common features of the conditions linked to uric acid production such as rheumatism was their periodicity—they were recurrent disorders. These developments sustained a widespread use of lithium in medicines and waters from the mid-nineteenth through the mid-twentieth century; 7 UP was first marketed in 1929 as a lithium-containing lemon and lime drink.

Lithium crept into psychiatric use through two portals. First, it could be combined with salts like bromides, which were then in regular use within mental hospitals as sedatives. This made lithium bromide a popular sedative for psychiatrists such as William Hammond in Bellevue in the 1870s and neurologists such as Weir Mitchell.

The second route of entry was through speculation that some nervous problems might also stem from excess uric acid production. Carl Lange, a Danish neurologist, noted that many of the neurological conditions he was seeing were accompanied by a periodic loss of vitality. The loss of vitality he described had features in common with what would later be diagnosed as vital depression, but was not then seen that way in that non-melancholic depression as we now know it essentially did not exist in the nineteenth century. As with many other neurologists of the day, Lange gave lithium, but picked out that it seemed to make some difference for periodic nervousness. Carl's brother Fritz Lange, an asylum superintendent in Denmark, also began using lithium and reported that it appeared to be helpful particularly in periodic melancholias.

There are two fascinating features to this nineteenth-century "discovery" of lithium for nervous disorders and its possible mood-stabilizing or prophylactic effect. First, despite apparently good results from the use of lithium, this use was eclipsed when ideas about a uric acid predisposition went out of favor at the start of the twentieth century. When the rationale for its use went, the use of lithium went with it even though there appeared to be results to favor its continuing use. The second point of interest is that after lithium's reappearance in the 1950s, trials of lithium for depression failed to offer any evidence that it might be useful.

Lithium drifted out of favor in psychiatry and the rest of medicine from the turn of the twentieth century, until finally it was banned by the FDA in 1949. In the late 1940s, it had come into vogue as a salt substitute on the back of research, indicating that salt might contribute to hypertension, but this use of lithium was linked to cardiac difficulties and led to an FDA ban.

But in 1949, John Cade in Melbourne also reported on lithium's beneficial effects for mania. In an article reported in the *Medical Journal of Australia*, Cade gave details of ten patients, three of whom had chronic mania, six of whom had episodic mania in one form or the other, and the tenth who was schizoaffective.[5] A great deal of how the lithium and indeed the bipolar disorder story has developed in the close to 60 years since this article arguably hinged on Cade's witting or unwitting rhetorical skill in framing his cases. In all the cases of mania, he reported lithium as producing a dramatic effect on the mental state of the patients within days, even in cases that had been chronically ill for up to five years. These benefits were lost when the treatment was discontinued but were retrieved on reinstatement of the treatment. Five of the vignettes end with the patient leaving hospital, taking up work, and seemingly unlikely to ever come back to hospital. In the case of the one schizoaffective patient, it appeared that his affective symptoms improved but his delusions and hallucinations remained. In the case of six schizophrenic patients, there was some settling of their disturbance but no real change in the core features of their illness. In the case of three depressed patients, there was no apparent response to lithium.

These Lazarus-like responses provoked the greatest interest in France. A series of articles from 1951 through 1955 reported beneficial effects in mania and in nonmanic states and hinted at better results when lithium use was maintained.[6] The most intriguing findings came from Teulié and colleagues who reported that when given in epilepsy, lithium had a beneficial effect on the difficulties of personality that were then often reported in epileptics.[7]

Despite all this activity, and the fact that by the mid-1950s publications on lithium in French outnumbered all other national contributions combined, Plichet, reviewing the use of lithium in France in June 1954 in *La Presse Médical*, said that Cade's article had little impact in France.[8] Lithium was a troublesome treatment that was contraindicated in depression, would not replace ECT (electroconvulsive therapy) for mania, but might have some benefit in prolonging recoveries induced by ECT, but at a cost of possible deaths if not monitored closely.

Interest in lithium died out in France but picked up in Denmark. In 1952, Erik Stromgren, professor of psychiatry in Aarhus in Denmark, a figure held in universally high esteem, drew the attention of Mogens Schou, a research associate at the hospital, to Cade's article. Schou recruited a group of 40 manic patients to a placebo control, double-blind trial of lithium. This 1954 study is a candidate for the award of the first randomized controlled trial (RCT) in psychiatry.[9] He reported that lithium could be of benefit for some manic patients. Later studies from England,[10] and the United States,[11] pointed in a similar direction.

Lithium, however, was a drug that was cheaply and easily produced. It was not a drug that could be patented. No pharmaceutical company stood to make much money out of it. Lithium was rarely featured in the major psychopharmacological meetings of the late 1950s or early 1960s that heralded the dawn of a new era in psychiatry, following on the advent of antipsychotics such as chlorpromazine and haloperidol and antidepressants such as the tricyclic, imipramine, and the monoamine oxidase inhibitors (MAOIs) phenelzine, and tranylcypromine.

Some indication of the precarious position into which lithium had slipped by this stage can be got from considering the proceedings of the first major international meeting of the psychopharmacological era held in Rome in 1958. This meeting convened all the big names from the United States, Europe, Japan, South America, and the rest of world psychiatry. In a set of proceedings that ran to 720 pages of relatively small and condensed print, there was only one short four-page piece on lithium—from Schou.[12]

A second psychopharmacology conference held in Basel in July 1960 again convened all the leading lights from world biological psychiatry. This gave rise to a set of proceedings that were over 520 pages long, in which there was no mention of lithium.

At the third major meeting of the new psychopharmacology era, held in Munich in September 1962, the proceedings ran to 600 pages. Lithium and Schou were given one page—the final page. Here Schou states:

Through its title and the communications so far given, this morning's discussion seems about to create the false historical myth that 1962 is the tenth anniversary of the psychopharmacological era. This, however, is neither true nor fair because in 1949 the Australian, Cade, discovered the therapeutic efficacy of lithium salts against manic phases of the manic depressive psychosis.

There may be a number of reasons for the unjustified neglect of this drug during the years. One is its rather narrow spectrum of indication, namely the typical manias and especially the chronic ones; but a high specificity of a drug ought not to detract from the appreciation of it. Furthermore, it is known that lithium may under certain extreme conditions produce kidney damage . . . but the main reason for the neglect of lithium may be quite simply that lithium salts are so inexpensive that no commercial interests are involved. This drug has therefore completely lacked the publicity that is invariably given to drugs of higher money earning capacity.

It is indeed conspicuous that lithium does not appear in any of the many general surveys, in spite of its therapeutic value being proved in a group of patients, which was resistant to most other therapies. This may conceivably be due to mere ignorance but such a suggestion is perhaps impolite. I would rather think that lithium is omitted from these schemes, because it is chemically completely unrelated to any of the other drugs used in psychiatry. I am therefore in complete agreement with what Dr Akimoto has just said. We must not let schemes and terminology, however beautiful and logically satisfying they may be, rule our thinking and obscure our observational powers. If, because it is easy or out of a desire for systematization, we adhere to a too categorical classification of drugs, we run the grave risk of distorting truth and hampering scientific progress.[13]

All this was about to change. Taking into account all possible articles on lithium from the chemical and physiological literature, through therapeutics, there had been 28 articles up to 1899. Thereafter the pace of production stepped up to reach 4 per year on average by 1949. There were 20 articles per year through the 1950s, stimulated in great part with concerns about toxicity and efforts to monitor physiological levels. In the 1960s, the rate increased first to 30 per year, then hitting 100 per year in 1966, and finally peaking at 200 per year around 1968/69—when lithium became embroiled in controversies about its mood-stabilizing properties.[14]

In the early 1960s, Schou was contacted from England by Toby Hartigan and from Denmark by Poul Baastrup, both suggesting that they thought lithium was beneficial in preventing recurrences of depression. After reading Schou's lithium trial, Baastrup began a trial of lithium at Vordingborg Hospital in 1957. He treated 56 patients with lithium, finding that "when lithium was given continuously to manic depressive patients who had been in hospital for years their condition improved so much that at least some of them could be moved from the closed wards to the open."[15]

As part of the trial he had conducted a follow-up examination on discharged patients. After discharge, these patients had been asked to stop taking lithium, but eight of the patients had continued to take lithium despite this advice and two of them had even given lithium to manic-depressive relatives. They claimed that continuous lithium treatment prevented psychotic relapse. Intrigued by the observation Baastrup looked back over the three years that these patients had been on lithium and looked at the rate at which they had been having episodes during these three years compared with their previous rate, and it appeared that the rate of new episodes was reduced compared with previously.[16]

At the same time, Hartigan gave lithium to a group of 20 patients in Britain, seven of whom had recurrent depressive episodes only:

Finally I have been experimenting with lithium on a group of seven patients with frequent recurrent depressions. There is little in the literature to suggest that depressive syndromes are improved by the drug and it is certainly not to be advocated during the acute depression episode but I have been using it as a prophylactic against further depressions in these patients and have had very promising results in five of them.[17]

Meanwhile, Schou had begun to think of lithium, but also other antidepressants such as imipramine, in a different light and in a publication in 1963, he proposed that lithium might be a normothymotic or "mood-normalizer."[18]

Schou had a brother who from the age of 20 suffered from repeated attacks of depression, which periodically made him unable to work:

> The attacks usually lasted some months, and then disappeared, but they reappeared again and again, year after year, inevitably. Then, about 14 years ago (mid-1960s), he was started on maintenance treatment with lithium, and since then he has not had a single depressive relapse. He still needs to take the medicine to keep the disease under control, but functionally he is a cured man. You will understand what such a change meant to himself and to his wife and children, and how much of a miracle it appeared to us in the family.[19]

Baastrup and Schou went back to investigate the clinical course of Danish manic-depressive patients, some of whom had now been on lithium for six-and-half years. In a study reported in the *Archives of General Psychiatry* in 1967,[20] they reported that the frequency of episodes in patients after lithium was much less than it had been before and that the total length of time spent in hospital was also much less. This paper and the articles by Hartigan and Baastrup pointing to a prophylactic benefit of lithium not only in classic bipolar disorder but also in recurrent unipolar illness lit a touch fuse which engaged most of the major personalities and provoked one of the most celebrated controversies in psychiatry.

The fuse was set at a meeting at Gottingen in 1966 at which Schou and Michael Shepherd participated. Based at the Institute of Psychiatry in London, Shepherd at the time was the doyen of randomized trials within psychiatry. He was the author of the first perspective parallel group for placebo-controlled trial in medicine, a trial that had demonstrated reserpine to be an effective antidepressant. He launched one of the first multi-centered trials of its kind in medicine, comparing ECT with imipramine, phenelzine, and placebo. This appeared to show that phenelzine, the most commonly prescribed MAOI antidepressant, didn't work; a result that all but finished off the MAOI group of drugs.

In Gottingen, Schou presented the results from Baastrup and Hartigan's work, indicating that lithium might have a prophylactic effect for recurrent mood disorders. The presentation could be viewed as cutting directly across the points made by Shepherd at the same meeting, who had argued that the judgment of the clinical investigator needed to be subject to the rigors of an impersonal clinical trial process that would establish whether or not a specific effect of treatment could be shown. Lots of people got well simply being seen. Did more people get well taking this treatment than got well simply by being seen and supported by their clinician? Shepherd was asked to comment on Schou's findings and the idea that lithium might be prophylactic. His response that these observations were interesting but remained to be confirmed relegated mood stabilization to the realm of fantasy.

After the exchanges, Schou's wife took issue with Shepherd for being too hard on her husband. Schou joined them and in the course of the conversation made the problem worse by revealing he had a brother who had, on the basis of Baastrup's results, begun taking lithium for a recurrent depressive disorder and that since taking lithium there had been no recurrences. This for Shepherd confirmed his suspicion that Schou was a "believer" in lithium prophylaxis. For him further studies were not needed.

Shepherd and Barry Blackwell responded to Schou and Baastrup's paper in *Lancet* under the heading of "Prophylactic Lithium: Another Therapeutic Myth?"[21] They noted that Cade's open study suggested that 80% of patients with mania might be expected to respond to lithium, whereas Schou's own controlled trial showed that only a third of manic patients responded. Generally, they reported that almost all authors, including early Australian proponents of lithium, had indicated that lithium was not a treatment for depression—that states of depression responded poorly to lithium—and that in the only controlled trial of lithium given for depression, which Schou himself had undertaken, the trial had been discontinued when only one out of twelve patients responded.

They noted that Baastrup's eleven patients with recurrent manic depressive illnesses who declared "they cannot do without lithium" were only "18% of the original 60 patient group, of whom some did not respond to therapy, others did not reappear." While Baastrup and Schou had compared the frequency of episodes before and after lithium in 88 patients from among 156 who had received lithium, Blackwell and Shepherd pointed out, most of these patients had been treated with ECT or antidepressant drugs in the first instance so it was not the lithium that had got them well. The problem here was "in view of the natural history of affective illnesses and their fragmentation by physical methods of treatment, several patients could have had single rather than recurrent illnesses."

They pointed out that if you took any group of patients with recurrent mood disorders and followed them up over a period of time you were likely to find similar results almost regardless of what they were treated with. And, in fact, Blackwell and Shepherd suggested that they had done just this with a group of patients from the Institute of Psychiatry and that the results when looked at this way suggested that almost anything including phenelzine was prophylactic for recurrent mood disorders when, in fact, Shepherd's 1965 trial had seemed to show that phenelzine simply wasn't an antidepressant. This suggested that Schou and Baastrup's patients had self-selected themselves to be responders to lithium, and that had they been randomized into a controlled trial the benefits would have not been as apparent.

This heavy-handed response engaged the psychopharmacological establishment of the day on the side of Schou and lithium, against what the proponents of psychopharmacology heard as the voice of therapeutic skepticism. None of the other drugs in psychiatry had been discovered by randomized controlled trials, it was noted, were Shepherd and Blackwell suggesting that these didn't work either? Shepherd and Blackwell's response was to see Schou and the proponents of lithium as enthusiasts whose repeated discoveries of elixirs of life were doing no one any favors.[22]

The upshot was a series of further controlled trials. In two of these, it appeared that the administration of lithium did indeed lead to a reduced frequency of further affective episodes.[23] But on the other hand, Shepherd and colleagues in further studies showed that amitriptyline had comparable prophylactic effects[24] and that lithium and amitriptyline produced apparently comparable prophylactic effects when given to patients with unipolar depressive disorders.[25]

This controversy dragged lithium out of obscurity and made it one of the stars in the psychotropic firmament. It also apparently established the idea that a drug might exert a prophylactic effect—might stabilize mood—although this language only came later. But in the process many things got turned inside out. The original boost to lithium's trajectory was its apparent specificity to mania, now it was moving into orbit on the basis of a quite different action. It stabilized rather than cured, but did it do so in any way other than the way many other psychotropic drugs did?

As all this was happening, a further study eroded the notion of lithium's specificity. Sheard looking at a prison population reported that lithium reduced levels of aggression.[26] This could be seen as an all but complete contradiction to Cade's original 1949 report. Alternatively, if lithium responsiveness defines bipolarity, the Sheard result could be reinterpreted as evidence that many patients with personality disorders had in fact unrecognized bipolar disorders. This essentially is what happened somewhat later when a new generation of mood stabilizers appeared.

One of the key developments came in the domain of research criteria. With the impact of lithium, the creators of diagnostic criteria recognized the importance for research of a distinction drawn in the 1960s between unipolar and bipolar depressions. This led to a recognition that not only were there bipolar I disorders, which involved hospitalization for mania, but there were also depressions that might be accompanied by what could be seen as mood elevations that did not require hospitalization—bipolar II disorders. Where bipolar I disorder, or folie circulaire, was a rare disorder with an incidence of 1/100,000 per year, bipolar II it was discovered might affect up to 5% of the population, depending on how one interpreted the affective instability in personality and other disorders. Mood stabilizers need no longer "lack the publicity that is invariably given to drugs of higher money earning capacity."

The Background: Valproic Acid

The origin of the drug that eventually created the mood-stabilizer category lies in the Second World War and efforts by German scientists to produce butter substitutes.[27] These efforts led to the synthesis of valproic acid. After the war valproic acid was used as a common diluent for other drugs. In 1963, George Carraz of the Laboratoire Berthier at Grenoble, when asked to test out a new product for possible anticonvulsant properties dissolved the new compound in valproic acid. Testing failed to show any correlation between different doses of the experimental compound and anticonvulsant activity, yet the mixture was anticonvulsant. Carraz realized that the anticonvulsant properties stemmed from valproic acid and titrating the dose of this demonstrated the issue conclusively.

Carraz synthesized sodium valproate (Depakine) and sodium valpromide (Depamide) derivatives of valproic acid. He had links with Pierre Lambert at Bassens Hospital in Rhône-Alpe and as a result the primary tests of the anticonvulsant properties of both valproate and valpromide took place in Bassens Hospital.[28]

At that point in time most large asylums in Europe had significant populations of epileptic patients—from 10 to 20%. This gave a ready population in which to try out a new anticonvulsant. Lambert initially found that when valpromide replaced other anticonvulsants, such as phenobarbitone, it appeared to have psychotropic in addition to neurotropic effects. As Lambert described it: "patients felt more themselves, the mental stickiness, viscosity that had sometimes been there on older agents, was less. We saw the disappearance of tendencies to depression, sometimes even a mild euphoria."[29]

Epilepsy was then linked to a distinctive set of personality problems. Epileptic patients were seen as importunate, manipulative, and viscous. These patients were frequently detained in hospital not because of their convulsions but because of the social disturbances they caused. They were thought to have impulse control disorders, which underlay their inability to adapt to normal social life. They were obsessional. On valpromide, these social disturbances and the importunate behavior of hospitalized epileptics appeared to change. Females were less likely to end up in conflicts, less likely to provoke others in their surroundings, and less likely to self-harm, leading Lambert to ask whether valpromide was anti-masochistic?

As lithium had fallen out of favor in France, there was a premium on finding effective treatments for manic-depressive disorder. The standard treatment involved antipsychotics such as chlorpromazine. The sedative properties of valpromide led to its use in combination with chlorpromazine for agitated and manic patients, just as phenobarbitone had been used. On recovery, patients appeared happier to continue valpromide than chlorpromazine. Altogether Lambert and colleagues studied the drug in approximately 250 patients and concluded that valpromide had distinct psychotropic effects that were of benefit in the treatment of both acute manic states and in the maintenance treatment of manic depressive illness—that it was what would now be called a mood stabilizer.[30]

This discovery of the psychotropic effects of valproate exactly parallels the discovery of the psychotropic effects of another anticonvulsant, carbamazepine, during the 1970s in Japan.[31] These joint discoveries pose the following question. The degree of control of convulsions is not significantly better now compared with before, but it is clear that epileptic patients do not end up in mental hospitals in a way that they did before. Is this because of a beneficial effect of these drugs on personality and general integration that has been all but uninvestigated? Is this beneficial effect what underpins mood stabilization?

Any debate there might have been on this issue, however, was cut short by Abbott Laboratories. Sodium valproate was an old molecule coming off patent, when it became more generally accepted that it might be useful in mood disorders. In an extraordinary example of how lax patent law had become by the 1980s—at least where pharmaceuticals were concerned—Abbott managed to secure a patent on semisodium valproate (Depakote) on the back of claims that it was more sparing to gastrointestinal function than sodium valproate. This led to clinical trials of the new compound and on to its licensing by regulators in America for the treatment of mania in 1994.

With the launch of Depakote, a new mania was born, for which a new term—mood stabilization—was needed. What mood stabilization might mean hinged loosely on an idea put forward by Bob Post at the NIMH that anticonvulsants worked in epilepsy by quenching the risk of further fits and that in the same way they might reduce the risk of any one episode of a mood disorder kindling the risk of further episodes. If this was the case, then potentially any anticonvulsant might be a mood stabilizer.

Back to the Future

The late 1950s and early 1960s was a time when there was a profusion of psychotropic agents used for a variety of indications. Compounds for the treatment of nervous problems, now written out of history, such as phenaglycodol, aspartate salts, buclizine, mephenoxalone, emylcamate, meparfynol, ectylurea, hydroxyphenamate, oxanamide, captodiame, chlormezanone, ethchlorvynol, glutethimide, benactyzine, deanol, flourished. Most of these vanished in the mid- to late 1960s in the wake of the thalidomide disaster, killed off by a drug evaluation project inaugurated by the FDA.[32] There were, in fact, probably a much greater number of drug classes available for clinical use in the late 1950s and early 1960s than are currently available now.

In addition to a greater variety of agents, there was also a much wider range of terms in use from the late 1950s through the 1960s to describe the effects of these drugs. These terms included sedatives, tranquilizers, calmatives, ataractics, thymoleptics, mood enhancers, mood stabilizers, mood normalizers, normothymotics, vegetative stabilizers, nootropics, psyche stabilizers, and hypnotics. Few of these designations have survived other than the term hypnotic.

The concept of a stabilizer seems almost inevitable when it comes to psychotropic drugs. The early psychotropics were either sedatives or stimulants. Sedatives, such as barbiturates, bromides, and chloral, were clearly aimed at damping down the extremes of behavior. But stimulants such as camphor and later strychnine were also aimed at restoring normality by increasing tone. These older ideas about restoring tone to some function were invoked to underpin the later use of stimulants such as dexamphetamine in states such as attention-deficit-hyperactivity disorder (ADHD). It was this restoration of tone that stabilized behavior.

It is therefore not surprising to see the word "stabilizing" used early on in the psychotropic era. And, in fact, the first use of the term mood stabilizer appears in the title of a 1960 article by Harry Litchfield, a pediatrician at Brooklyn Women's and Rockaway Beach Hospital, on aminophenylpyridone, also known as aminophenidone,[33] one of the many meprobamate-benzodiazepine-barbiturate-related compounds that appeared in the late 1950s and early 1960s. Others using this drug referred to it as an emotion stabilizer[34] and either a calmative or mood stabilizer.[35]

Aminophenidone was launched as Dornwal in April 1960 by the pharmaceutical company Wallace and Tiernan, through a retail division called Maltbie Laboratories. The advertisements suggested that it was effective for tension and for anxiety states and would avoid the problems of habituation and sedation linked to barbiturates and to the best-selling replacement for barbiturates—meprobamate. This was a niche that Librium, being launched at almost exactly the same time as Dornwal, was to colonize and take over all but completely.

A few years earlier, the overlapping introductions of meprobamate, reserpine, and chlorpromazine (Thorazine) had given rise to the notion of a tranquilizer. Chlorpromazine and reserpine were effective in a range of nervous states without producing sedation. This was caught in later references to the effects of chlorpromazine as equivalent to those of a chemical lobotomy. And it was a consideration of these effects of reserpine that led FF Yonkman of Ciba in 1953 to come up with a new term tranquilizer.[36] Reserpine didn't sedate the way that older sedatives did and didn't stimulate but rather appeared to "chill" the personality in ways that were quite different from anything that had been seen before. This was what might now be termed an anxiolytic effect, but in the 1950s, the idea that a psychotropic drug might be anxiolytic in ways that were quite independent from any sedative or stimulant effect it had had still not been born. The notion of a tranquilizer was the first step down this path.

Although both clearly differed from older sedatives, the effects of chlorpromazine and reserpine on the one hand and those of meprobamate and the benzodiazepines on the other hand were quite different. This difference split the tranquilizer group into the major tranquilizers, chlorpromazine and reserpine, which were used primarily for psychotic disorders and very quickly transmuted into antipsychotics, and the minor tranquilizers, a group containing meprobamate and the benzodiazepines. The term tranquilizer supplanted notions of sedatives and became the dominant term by which psychotropic drugs were described in the West through the 1980s, when the benzodiazepines became embroiled in a crisis regarding their potential to cause dependence. As a consequence of this crisis, the notion of a tranquilizer became a forbidden notion. Where once it stood for an escape from the hazards of sedatives, now it conjured up a vision of dependence, and it too had to be replaced. The replacements were two new terms—anxiolytic and mood stabilizer.

In Japan, during this period, efforts to translate the term tranquilizer produced the notion of a seishin-antai-zai, literally a psyche-stabilizer. But where psyche-stabilizer remained continuously in use in Japan, the term mood stabilizer, which essentially seems to have functioned in the same way when applied to Dornwal in 1960, vanished.

Maltbie was trying both to edge into what Miltown had made a fabulously lucrative market and to differentiate Dornwal from Miltown. One way to achieve both was to rebrand the new group of drugs. SmithKline Beecham, as a late entrant into the serotonin reuptake inhibiting class, did just this with extraordinary success 30 years later when they christened Paxil the selective serotonin reuptake inhibitor, the SSRI. Rebranding even Prozac in SmithKline colors played a part in ultimately overtaking it as the best-selling SSRI.

It's difficult to know now whether Wallace and Tiernan's efforts in this area reflect as deliberate a marketing strategy by the company as this or whether what is involved is a transitional phase, in which what might look like solid concepts pass through periods of greater fluidity than seems apparent retrospectively when awareness of the competing possibilities has vanished.

Most users of the concept of an antidepressant now, for instance, probably think its meaning is pretty clear and it has always meant something close to what it now means. But in the 1960s it too probably meant something close to a mood stabilizer, whereas quite unaware of this the advocates of pharmacological mood stabilization now argue against using antidepressants in people with a bipolar disorder. Take advertisements from 1963/64 for Eli Lilly's new antidepressant nortriptyline, traded as Aventyl in Britain for instance, which referred to a new wide spectrum mood enhancer:

> When the patient is overwhelmed by normal every day stresses Aventyl often brings prompt relief. Even in the more serious behaviour disturbances—whether marked by depression or anxiety, both Aventyl with its unusually wide spectrum of action, broadens the doctor's power to relieve the troubled mind swiftly.
>
> "Aventyl" has an inherently beneficial effect on anxiety and hostility—it is not a combination of an antidepressant with a tranquilliser, but a *single entity* going beyond mere antidepressant action as such. Four out of five patients with symptoms referable to anxiety and depression (including psychosomatic disturbances) are likely to respond favourably to "Aventyl" many of them within a week.

The circumstances of Dornwal's first use probably had something to do with the disappearance of the term mood stabilizer for a generation. Early in its career, Dornwal was linked to blood cell suppression. Reports trickled into the company of fatalities and other serious events. The medical director of Wallace-Tiernan, Charles Hough, was aware of these. The company later claimed that it had a duty to evaluate these reports fully before forwarding them to the FDA.[37] Frances Kelsey of the FDA, then sitting on a licensing application for thalidomide, was made aware through a series of accidental conversations of a possible problem. In November 1961, Dornwal was removed from the market. Wallace-Tiernan were subsequently indicted. After a period of huffing and puffing, they pleaded no contest.

One of the early evaluations of Dornwal's use as an agent for nervous disorders came from McGill's Hazim Azima, whose article reporting his findings[38] was ironically published a few pages earlier in the *American Journal of Psychiatry* than another Azima article evaluating a new hypnotic—thalidomide.[39] And it was Kelsey, perhaps sensitized by her experience with Dornwal, who subsequently drew attention to the risks of thalidomide.

Dornwal's success might have established the concept of a mood stabilizer in the early 1960s, but if it had done so this term would have had an entirely different set of connotations to those clustering around the 1990s incarnation of the term, even though the mood stabilizers of the 1960s, the benzodiazepines, are all anticonvulsant. When the term reemerges in the 1990s, all awareness of the dark star under which it had been born in the 1960s had vanished.

There are a number of oddities in Chouinard's use of the term mood stabilizer in his 1985 letter to the *American Journal of Psychiatry*.[40] On the one hand, the letter suggests a concept in widespread use, but there are no references to any prior use of the term or any efforts to justify its use. On the other hand, the idea that an estrogen-progesterone combination might be a member of some recognized class was so unusual at the time that using the term in this context almost necessarily implied that this usage was groping toward something new. The next use of the term, also from Chouinard, referred to magnesium aspartate as a mood stabilizer,[41] another compound that is not traditionally seen as a psychotropic agent.

Reviewing the state of play in 1987, Leo Hollister, one of the grandees of American psychopharmacology, noted that there were some advocates of the term mood stabilizer but that it had not yet come into use and in Hollister's opinion the concept hopefully wouldn't take off.[42] He was doubtful that the prophylaxis that the term implied was possible. Hollister's was the only entry for mood stabilizer in the index of this large 1,780-page, double column, small print book with quarto-sized pages. In the early 1990s, most books on psychotropic drugs for professionals[43] or patients[44] still had no mention of the term mood stabilizer.

During the period from 1985 to 1994, further uses of the term that appear in Medline suggest that the concept of a mood stabilizer had not settled down. Jonathan Cole from Harvard applied the term to flupenthixol, an antipsychotic then widely used in Europe but not available in the U.S. market.[45] This agent had been cultivated in Europe by Lundbeck, as an antipsychotic in high doses but an antidepressant or anxiolytic in low doses. Cole and his group also referred to clozapine as a mood stabilizer when it was relaunched in the United States at the end of the 1980s.[46] Finally, Les Grinspoon and Jim Bakalar applied the term to cannabis.[47]

All these uses of the term were a long way from branding of the mood stabilizer that Abbott engineered after the 1994 launch of Depakote. On the coattails of Depakote, the term was transformed from its former meaning of something close to a tranquilizer or psyche-stabilizer into a term that meant a nontoxic equivalent of lithium. Mood stabilizers were now something that prevented further episodes of a mood disorder by acting on physiological disturbances in some mood center rather than agents, which were in some sense tranquilizers or anxiolytics.

Based on Post's idea that anticonvulsants quenched the risk of further episodes of mood disorders, it was assumed by many that all anticonvulsants would in fact be mood stabilizers and most anticonvulsants began to be used off-label for just this purpose. Encouraged by the placement of ghostwritten articles in major journals, Pfizer's gabapentin, which had been making about $300 million a year for epilepsy, for instance, soared to a billion dollar blockbuster in great part because of its use for mood disorders.

Antipsychotics such as olanzapine (Zyprexa), a drug with an extraordinary propensity to cause diabetes, and the highest suicide rate in clinical trial history, were rushed into clinical trials of mania and then maintenance trials for bipolar disorder and emerged as mood stabilizers and in short order were being given to children aged one and up—even though no one in their right mind would have considered giving chlorpromazine or haloperidol to such children.

A book, *The Bipolar Child*, appearing in 2000 sold 70,000 hardback copies in six months.[48] America and psychopharmacology have always had their enthusiasms, and America it seemed was now in the grip of a full-blown psychopharmacological enthusiasm. Problems could be solved if only they were detected at the earliest possible opportunity. The only reason mood stabilizers, sometimes in combinations of four or five or six different drugs together, didn't always work later in life was that the disorder had taken too deep a root and the answer was to start treatment at the first hints of affective instability, no matter how young the child might be.

All of this was a long way from Falret's 1851 delineation of folie circulaire. All of this proved too much for Carraz, who became reclusive and paranoid. He made regular claims that he had unearthed new breakthrough molecules, but that a range of forces were conspiring to deny him due credit.[49] All of this was essentially without a basis in clinical trial data. Even the foundational idea underpinning the bipolar bubble—that anticonvulsants might quench the risk of further episodes of a recurrent disorder—was wrong; anticonvulsants do not quench the risk of further episodes even in people with epilepsy.[50] All of which qualifies this episode in the medicating of America as a folly, perhaps the supreme folly of the era.

NOTES

1. Guy Chouinard, Susanne Steinberg, Warren Steiner, "Estrogen-progesterone combination: another mood stabilizer?" *American Journal of Psychiatry* 144 (1985): 826.

2. Gary Sachs, "Bipolar Mood Disorder: Practical strategies for acute and maintenance phase treatment." *Journal of Clinical Psychopharmacology* 16(1996): 32s–47s; Charles L Bowden, "New concepts in mood stabilization: Evidence for the effectiveness of Valproate and Lamotrigine," *Neuropsychopharmacology* 19 (1998): 194–199; S Nassir Ghaemi, "On defining 'mood stabilizer.'" *Bipolar Disorder* 3(2001): 154–158.

3. Michael Berk, Seetal Dodd, "Bipolar II Disorder: a review." *Bipolar Disorder* 7 (2005): 11–24.

4. Jennifer Harris, "The increased diagnosis of "juvenile bipolar disorder": what are we treating?" *Psychiatric Services* 56 (2005): 529–531.

5. John FJ Cade, "Lithium salts in the treatment of psychotic excitement." *Medical Journal of Australia* 2 (1949): 349–353; See also John FJ Cade, "Lithium in psychiatry: historical origins and present position." *Australian and New Zealand Journal of Psychiatry* 1(1967):61–62.

6. M Despinoy, J Romeuf de, "Emploi des sels de lithlium en thérapeutique clinique." *Compters Rendus du Congrès des Médécins Alienistes et Neurologistes de Langue Francaise* (1951): 509–15; R Reyss-Brion, J Grambert, "Essai de traitment des états d'excitation psychotique par le citrate de lithium." *Journal de Médécine de Lyon* 32(1951): 985–89; M Deschamps, M Denis, "Premiers resultats du traitement des états d'excitation maniaque par les sels de lithium." *L'Avenir Medical* 49 (1952): 673–79; N Duc, H Maurel, "Le traitement des états d'agitation psychomotrice par le lithium." *Le Concours Médical* 75 (1953): 1817–1820; J Carbère, M Pochard, "Le citrate de lithium dans le traitement des syndromes d'excitation psychomotrice." *Annales Médico-Psychologiques* 112(1954): 566–72; M Teulié, M Follin, M Begoin, "Etude de l'action des sels de lithium dans états d'excitation psychomotrice." *Encéphale* 44 (1955): 266–85; J Oulès, R Soubrie, P Salles, "À propos du traitement des crises de manie par les sels de lithium." *Comptes Rendus du Congrès des Médécins Alienistes et Neurologistes de Langue Francaise* (1955): 570–73; J Oulès, "'Discussion,'" *Annales Médico-Psychologiques* 113 (1955):

679–681; P Sivadon, P Chanoit, "L'emploi du lithium dans l'agitation psychomotrice à propos d'une experience clinique." *Annales Médico-Psychologiques* 133 (1955): 790–96; C Maissin, "'Le Traitement de la Manie par le Citrate de Lithium." *Thesis presented for the Doctorate of Medicine in the Faculty of Medicine,* Paris (1955): 48.

7. M Teulié, M Follin, M Begoin, "Etude de l'action des sels de lithium dans états d'excitation psychomotrice." *Encéphale* 44 (1955): 266–85.

8. A Plichet, "Le traitement des états maniaques par les sels de lithium." *Presse Médical* 62 (1954): 869–70.

9. Mogens Schou, Niels Juel-Nielsen, Erik Stromgren, Holger Voldby, "The treatment of manic psychoses by the administration of lithium salts." *Journal of Neurology, Neurosurgery and Psychiatry* 17 (1954): 250–260; Mogens Schou, "Phases in the development of lithium treatment in psychiatry." In *The Neurosciences: Paths of Discovery II,* ed Fred Samson, George Adelman (Birkhauser, Boston, 1992), 148–166.

10. Samuel Gershon, Arthur Yuwiler, "Lithium iron: a specific pharmacological approach to the treatment of mania." *Journal of Neuropsychiatry* 1 (1960): 229–241.

11. Ronald Maggs, "The treatment of manic illness with lithium carbonate." *British Journal of Psychiatry* 109 (1963): 56–65.

12. Mogens Schou, "Therapeutic and toxic properties of lithium." In Philip Bradley, Pierre Deniker, Corneille Radouco-Thomas (eds.) *Proceedings of the First International Congress of Neuropharmacology,* Rome, September 1958 (Elsevier, and Sudam, 1959), 687–690.

13. Mogens Schou, *Proceedings of the Third CINP Congress* (Elsevier, Amsterdam, 1962), 600.

14. F Neil Johnson, John FJ Cade, "The historical background to lithium research and therapy." In F Neil Johnson (ed) *Lithium Research and Therapy* (London, Academic Press, 1975), 9–21.

15. Cited in F Neil Johnson, *The History of Lithium Therapy* (Macmillan, London, 1984), 71.

16. Poul C Baastrup, "The use of lithium in manic depressive psychosis." *Comprehensive Psychiatry* 5 (1964): 396–408.

17. Toby GP Hartigan, "Experiences of treatment with lithium salts." *Journal of Mental Science* (1961): S49.

18. Mogens Schou, "Normothymotics, "mood-normalizers": are lithium and imipramine drugs specific for affective disorders?" *British Journal of Psychiatry* 109 (1963): 803–809.

19. Mogens Schou, *Address on receiving an honorary doctorate* (University of Aix-Marseilles, 29th October 1981).

20. Poul C Baastrup, Mogens Schou, "Lithium as a prophylactic agent: Its effect against recurrent depression and manic depressive psychosis." *Archives of General Psychiatry* 16 (1967): 162–172.

21. Barry Blackwell, Michael Shepherd, "Prophylactic lithium: Another therapeutic myth? An examination of the evidence to date." *Lancet* (1968): 968–970.

22. Barry Blackwell, "Letter," *Journal of the American Medical Association* 125 (1969): 1131; Nathan S Kline, "Letter," *Journal of the American Medical Association* 125 (1969): 1131–1132.

23. Alec Coppen, Ramon Noguera, John Bailey, Bruce Burns, MS Swami, Edward Hare, Raymond Gardner, Ronald Maggs, "Prophylactic lithium in affective disorders. Controlled trial." *Lancet* (1971): 275–279.

24. Richard S Mindham, C Howland, Michael Shepherd, "Continuation therapy with tricyclic antidepressants in depressive illness." *Lancet* (1972): 854–855.

25. A Ian Glen, Al Johnson, Michael Shepherd, "Continuation therapy with lithium and amitriptyline in unipolar depressive illness: a randomised double blind controlled trial." *Psychological Medicine* 14 (1984): 37–50.

26. Michael H Sheard, "Effect of lithium on human aggression." *Nature* 230 (1971): 113–114.

27. JW Meijer, H Meinardi, CD Binnie, "The development of antiepileptic drugs." In Michael J Parnham, Jacques Bruinvels (eds.) *Discoveries in Pharmacology Vol 1* (Elsevier, Amsterdam, 1983).

28. Georg Carraz, S Lebreton, M Boitard, Sergio Borselli, J Bonnin, "À propos de deux nouveaux anti-epileptiques de la série n-dipropylacetique." *Encéphale* (1965): 458–465; Pierre Lambert, Sergio Borselli, G Marcou, M Bouchardy, Georg Carraz, "Propriétés neuro-psychotropes du Depamide: action psychique chez les epileptiques et les malades presentant des troubles caracteriels." *CR Congr de Psychiatrie et de Neurologie de langue Française, Masson, Paris* (1966): 1034–1039; Pierre Lambert, Sergio Borselli, J Midenet, C Baudrand, G Marcou, M Bouchardy, "L'action favorable du Depamide sur l'évolution à long terme des psychoses maniaco-depressives." *CR Congr de Psychiatrie et de Neurologie de langue Française, Masson, Paris* (1968): 489–495.

29. Comité Lyonnais de Recherches Thérapeutiques en Psychiatrie, "The birth of psychopharmacotherapy: explorations in a new world—1952–1968." In David Healy (ed) *The Psychopharmacologists Vol 3* (Arnold, London, 2000), 1–54.

30. Pierre A Lambert, Georg Carraz, Sergio Borselli, M Bouchardy, "Le dipropylacetamide dans le traitement de la psychose maniaco-depressive." *Encéphale* (1975): 25–31.

31. Margaret Harris, Summit Chandran, Nabonita Chakroborty, David Healy, "Mood stabilizers: the archaeology of the concept." *Bipolar Disorders* 5 (2003): 446–452.

32. Edward Shorter, "Looking backwards: a possible new path for drug discovery in psychopharmacology." *Nature Reviews. Drug Discovery, Perspective 1* (2002): 1003–1006.

33. Harry Litchfield, "Aminophenylpyridone, a new mood stabilizing drug." *Archives of Paediatrics* 77 (1960): 133–137.

34. AL Cantelmo, "Clinical evaluation of aminophenylpyridone (dornwal): a new drug for stabilizing emotional behavior." *Current Therapeutic Research and Clinical Experience* 2 (1960): 72–75.

35. Leo Cass, Willem Frederik, Jose Teodoro, "Evaluation of calmative agents." *American Practitioner & Digest of Treatment* (1960): 285–288.

36. Hugo Bein, "Biological research in the pharmaceutical industry with reserpine." In Frank J Ayd, B Blackwell (eds) *Discoveries in Biological Psychiatry* (Lippincott, Phila Pa.,1970), 142–152.

37. Morton Mintz, *By Prescription Only* (Houghton-Mifflin, Boston, 1967).

38. Hazim Azima, Dorothy Arthurs, A Silver, "The effect of aminophenidone in anxiety states, a multi-blind study." *American Journal of Psychiatry* 118 (1961): 159–160.

39. Hazim Azima, Dorothy Arthurs, "A controlled trial of thalidomide, a new hypnotic agent." *American Journal of Psychiatry* 118 (1961): 554–555.

40. Guy Chouinard, Susanne Steinberg, Warren Steiner, "Estrogen-progesterone combination: another mood stabilizer?" *American Journal of Psychiatry* 144 (1985): 826–827.

41. Guy Chouinard, Linda Beauclair, Rita Geiser, Pierre Etienne, "A pilot study of magnesium aspartate hydrochloride (Magnesiocard) as a mood stabilizer for rapid cycling bipolar affective disorder patients." *Progress in Neuro-Psychopharmacology & Biological Psychiatry* 14 (1990): 171–80.

42. Leo Hollister, In Herbert Meltzer et al (eds) *Neuropsychopharmacology. The Third Generation of Progress* (Raven Press, New York, 1987), 35.

43. Brian Leonard, *Fundamentals of Psychopharmacology* (J Wiley & Sons, Bristol, 1992); Frank J Ayd Jr (eds) (Lexicon of Psychiatry, 1996).

44. Jack Gorman, *The Essential Guide to Psychiatric Drugs* (St Martin's Press, New York, 1992).

45. Amanda Gruber, Jonathan Cole, "Antidepressant effects of flupenthixol." *Pharmacotherapy* 11 (1991): 450–459.

46. Carlos Zarate, Mauricio Tohen, Michael Banov, Michelle Weiss, Jonathan Cole, "Is clozapine a mood stabilizer?" *Journal of Clinical Psychiatry* 56 (1995): 108–12.

47. Lester Grinspoon, James Bakalar, "The use of cannabis as a mood stabilizer in bipolar disorder: anecdotal evidence and the need for clinical research." *Journal of Psychoactive Drugs* 30 (1995): 171–77.

48. Demitri Papolos, Janice Papolos, *The Bipolar Child* (Random House, New York, 2000).

49. Comité Lyonnais de Recherches Thérapeutiques en Psychiatrie, "The birth of psychopharmacotherapy: explorations in a new world—1952–1968." In Healy, *The Psychopharmacologists Vol 3*.

50. Anthony Marson, Ann Jacoby, Anthony Johnson, Lois Kim, Carol Gamble, David Chadwick, "Immediate versus deferred antiepileptic drug treatment for early epilepsy and single seizures: a randomised controlled trial." *Lancet* 365 (2005): 2007–2013.

3

HORMONE REPLACEMENT

"Educate Yourself"

Consumer Information about Menopause and Hormone Replacement Therapy

Elizabeth Siegel Watkins

In the 1990s, Wyeth-Ayerst dominated the market for menopausal and postmenopausal hormone replacement therapy (HRT) in the United States with its two products, Premarin (conjugated estrogens, introduced in 1942) and Prempro (a combination estrogen-progestin pill, introduced in 1995). Together, these two drugs accounted for two-thirds of all prescriptions for HRT. From 1991 to 1999, Premarin enjoyed enormous success as the best-selling drug in America, occupying either the first or second spot on the list of most popular prescriptions. In 2001, although it had been knocked down to number three because of the popularity of Parke-Davis's blockbuster statin, Lipitor, Premarin still sold an impressive 38 million prescriptions, while Prempro sold almost 19 million, out of an astonishing total of 91 million for all HRT drugs. But in 2003 prescriptions for Premarin fell by 33 percent, Prempro's plummeted by 66 percent, and the total number for HRT declined to less than 57 million in all. In 2004, Premarin ranked only 31st in prescriptions written that year.[1] What happened to cause these drugs to fall out of favor so quickly? In the summer of 2002, the federally funded Women's Health Initiative study—a randomized double-blind controlled trial with 16,000 participants—reported shocking results: HRT users were at increased risk for heart disease, strokes, blood clots, and breast cancer. The story made the front pages of newspapers across the country, because it affected the lives of millions of American women.[2]

Since the 1960s, doctors had been prescribing estrogen for their older female patients both to treat short-term menopausal symptoms (such as hot flashes and night sweats) and to prevent long-term diseases allegedly caused by the reduced estrogen production of the postmenopausal years (such as osteoporosis and heart disease). Over the years, the rationale for prescribing hormone therapy changed in response to new scientific data about female sex hormones and to shifts in cultural inclinations regarding the roles of medicine and older women in society. In the 1960s and early 1970s, this class of drugs offered promises of continued femininity to aging women in a culture that celebrated youth. In the mid-1970s, faced with new medical evidence about the potential danger of long-term estrogen use and new feminist critiques of sexist assumptions about women in America, medical attitudes toward hormone therapy became more cautious. These attitudes changed yet again in the 1980s in response to medical reports that the addition of a second hormone, progestin, to estrogen in replacement therapy would thwart the development of endometrial cancer and that estrogen could confer protection against the development of osteoporosis and heart disease.[3] In the 1990s, between one in four and one in six American women over the age of 50 was taking estrogen.[4] By this time, many women had imbibed the message—from their doctors, from pharmaceutical promotional materials, from the popular media—that hormone replacement was a positive, proactive therapy that would keep them looking and feeling young and healthy. The news from the Women's Health Initiative flew in the face of accepted medical practice in the United States and threw the consumers of monthly prescriptions of this received wisdom into a state of panic. Although abundant information existed on the subject of HRT, none of it offered clear advice about what to do in the wake of this latest medical finding.

For more than 40 years, women had looked to a variety of sources to learn about health issues during and after menopause. Some went to their doctors for advice and relief; many more talked to friends or turned to the newsstand for published information. National surveys give some indication of the relative importance of different health education sources. In 1978, General Mills sponsored a national study of American families which found that 25 percent of respondents named popular magazines and newspapers and 30 percent named television as their main media of health information.[5] The Midlife Women's Health Survey in 1990 found that 33 percent of respondents cited magazines and books as their main resources for learning about menopause; just half that number (16%) turned to physicians and other health care professionals for information.[6] This study also reported television as a major source of information about menopause, and not just news broadcasts and educational programs; one woman said she learned about menopause from Edith Bunker on "All in the Family."[7] In 1991, *McCall's*, a women's magazine, surveyed its readers about menopause and found that 83 percent gleaned their information from magazines, newspapers, books, and television; only half that number (44%) included doctors on their list. Forty percent identified the media as their best educational resource; doctors were credited by just 18 percent.[8] Since the 1970s, the authoritative voice of the physician has increasingly had to compete with other opinions, advice, and information readily accessible to women.

These surveys confirm a significant role for popular media in disseminating information about health and, more specifically, menopause. The information about menopause and hormone replacement therapy, and the way that information has been presented, has changed over the course of the past several decades, in response not only to developments in medical knowledge, but also to two important social and demographic shifts: the women's movement and the aging of the baby boom generation. Prior to the 1970s, most of the information about menopause and aging available to older women in books and magazine articles was provided by doctors and drug companies. Journalists reported what they heard at medical conferences or what they read in medical journals. Oftentimes the authors were physicians themselves, reporting on their own personal experiences with menopausal patients. As a result of the women's movement (with contributions from the consumers' movement and the patients' rights movement as well), the equation that knowledge equals power was applied to the doctor-patient relationship. All patients, not just women, were encouraged to become better informed, so that they could be active participants. The women's health movement of the 1970s, along with a growing health consciousness among middle-class Americans at that time, contributed to an expansion in the quantity and variety of health information sources on menopause and aging. An even greater upsurge in public discourse on these topics came in the 1990s, as the first baby boomers reached menopause. This large demographic group of women has participated in the building of a community of active, informed health care consumers who have influenced both the provision of and access to information about menopause and aging.

It is important to note that HRT is not a life-saving medication, nor does it cure any disease. Although some physicians have considered menopause to be an estrogen-deficiency disease akin to hypothyroidism or diabetes, it is not an illness or disorder, but rather a normal stage in the female life cycle, completing a woman's years of fertility that began with menarche in her adolescence. HRT has been prescribed to improve one's quality of life, by easing menopausal discomfort or by (allegedly) preventing the development of age-related disease. As consumers of this optional therapy, women became increasingly involved in choosing whether or not to take it, rather than acting only on the advice of their doctors. In order to participate in such medical decision-making, they needed access to specialist knowledge. This chapter explores the changes in the composition and dissemination of popular information on hormone replacement therapy for consumers over the past forty years. This sort of health education incorporated not only the translation of data from the language of science into plain English, but also advice about who should use hormone replacement and when and why. These recommendations were not solely based on the current state of medical knowledge, which was often unclear, but also reflected cultural conceptions of medicine, medicalization, and female aging, which were also fraught with ambiguity.

The Discovery and Early Use of Estrogen

In 1929, Edgar Allen and Edward Doisy collaborated in the discovery and isolation of the female sex hormone, estrogen. Within a few years, several North American and European pharmaceutical companies had capitalized on this scientific development and brought estrogen products onto the market. In 1942, Ayerst Laboratories, a Canadian drug company, sought and won approval from the FDA to market an orally administered form, called conjugated estrogens, derived from pregnant mares' urine (the brand name Premarin is derived from this source). Premarin would come to dominate the hormone replacement market, accounting for 70 percent of estrogen prescriptions for menopausal and postmenopausal use by the 1970s.[9]

Initially, physicians prescribed estrogen as a short-term palliative for menopause-specific complaints. Mainstream advice endorsed a three-tiered approach to treating patients suffering from menopausal symptoms: first physician counseling, then sedatives, and finally, if all else failed, a limited course of hormone replacement. Historian Judith Houck has characterized this medical encounter between doctors and middle-aged patients as respectful of women's experiences, because "physicians treated women as a whole and menopause as a complex process, only part of which was rooted in women's glands."[10] This situation began to change in the 1950s, when some physicians became convinced that the problem of aging lay squarely in women's declining ovaries and began to promote the long-term use of hormones to combat a variety of alleged menopause-related conditions, namely, osteoporosis, cardiovascular disease, and mental senility. Although these doctors' ideas fell outside the medical consensus of the time, they articulated a position toward long-term hormone therapy that would become increasingly popular in the next decade.[11]

Selling Long-Term Estrogen Replacement to Women in the 1960s

As more and more physicians began to prescribe estrogen for their female patients in the 1960s and early 1970s, popular magazines warmed to the possibility of long-term hormone replacement. Some journalists parroted the notion that menopause was an estrogen deficiency disease; others, although unwilling to let go of the idea that menopause was a natural process, presented menopause in a much more negative light than in previous years. However, there was a solution at hand, and many of these writers presented estrogen as something close to a miracle drug. One doctor, writing in the *Ladies' Home Journal*, asserted:

> Despite baseless fears, estrogen is usually safe and effective in relieving the distressing symptoms of menopause. It is perfectly natural for women to wish to slow up the aging process and to remain more attractive. They don't hesitate to use contact lenses for failing eyesight, color rinses for drab-looking hair or caps for their teeth. Then why should they put up with the discomforts that afflict about half of them in middle age, when the menopause begins?[12]

Readers of this article were assured of the safety of this drug and encouraged to avail themselves of its beneficial effects. Another article in *McCall's* a few months later called "Pills to Keep Women Young" relied on "eight distinguished doctors who have done important work with estrogen [to] tell of its exciting results."[13] By the early 1970s, hormone replacement therapy had become incorporated into fashion and beauty magazines' recommendations for the over-40 crowd. *Harper's Bazaar* asserted in its "Over-40 Guide on Health, Looks, Sex" that "prevalent medical opinion is that the safety and benefits of estrogen therapy have been convincingly demonstrated."[14] The education available to readers of popular magazines prior to 1976 relied heavily on doctors as experts, promoting estrogen as a panacea for the woes of menopause and aging.[15]

Not all articles in popular periodicals were unreservedly positive. *Reader's Digest*, for example, cautioned its audience to exercise restraint in the medical management of menopause and postmenopause, telling readers that "No pill can make one young again. Nor can a pill make one feminine . . . nor is it a cure-all for the stresses and strains of a woman's life."[16] The author (in an article condensed from *U.S. Lady*) downplayed the number of women who truly needed estrogen at menopause (about 15 percent) and who were hormone deficient after menopause (about 25 percent). She abhorred the recent rush to estrogenate, blaming "enthusiastic articles and sensational advertisements" for inveigling women into believing in this mythical fountain of youth.

This author quoted the *Medical Letter* (a nonprofit publication that reviewed medications and therapies) as a respected and objective arbiter of medical claims, as did the author of an article in *Time* on the subject of estrogen. The *Medical Letter* recognized that estrogen could relieve immediate symptoms of menopause (such as hot flashes), and it conceded that the hormone played some role in treating osteoporosis (but not necessarily in its prevention); however, it rejected as premature the assertions that estrogen "promotes either the appearance or the feeling of youthfulness, or prevents or lessens the complications of arteriosclerosis," and recommended against the prescription of hormone replacement for all women.[17] The mass circulation magazines appealed to the authority of the specialist journal in an attempt to temper what they perceived to be a potentially dangerous fad.

By the mid-1970s, it appeared that the fad was not going to fade away. To assess the prevalence of hormone replacement therapy in medical practice, *McCall's* sent questionnaires to gynecologists, two hundred of whom responded. These participants were not a representative sample of American gynecologists: they were all women over the age of 40, "with a double interest in hormone therapy."[18] Both the results of the survey and the manner in which they were reported in this popular women's monthly expose a snapshot of the standing of estrogen in the eyes of doctors and magazine writers in 1975.

Among this group of female physicians, estrogen was, in a word, popular. Ninety-eight percent of the gynecologists had prescribed HRT for patients, and 58 percent of those who were menopausal used it themselves. Eighty-four percent believed estrogen alleviated psychological problems, such as depression and anxiety, and 51 percent recommended HRT as a preventive against osteoporosis and atherosclerosis. The final analysis, however, reflected some ambivalence within the medical profession about the definition and treatment of menopause. Was menopause an estrogen deficiency disease? Some said yes, others said no, and ultimately "the decision whether to use replacement hormones still remains one for an individual woman and her doctor to make."[19]

Thus, in leafing through their favorite women's monthly or news weekly, women learned about the existence of estrogen and the claims made for its anti-aging effects, but they received confusing and sometimes conflicting recommendations about the wisdom of long-term therapy. Those women who took magazine writers' advice and went to see their doctors encountered another source of information in the waiting room: pharmaceutical brochures for estrogen products.

Drug companies promoted the medical management of menopause through "informational" pamphlets, which were distributed to doctors' offices. Women received a one-sided message from these booklets: menopause, although a natural transition, caused unpleasant symptoms, such as "physical and emotional turmoil, sagging and flabbiness of the breasts, tendency to put on ugly weight, . . . embarrassingly copious perspiration, excruciating headaches, causeless crying spells, . . . senseless, addle-headed anxiety."[20] The pamphlets went on to offer a solution to these miseries: hormone replacement therapy, which, according to these guides, could safely be taken for the rest of one's life.

Within a culture that valued youth as the standard for beauty, particularly for females, women sought ways to thwart the aging process. Hormone therapy was one more weapon in an arsenal that included vitamins, hair dyes, and cosmetics. As Susan Sontag articulated in 1972:

> Women have a more intimate relationship to aging than men do, simply because one of the accepted "women's" occupations is taking pains to keep one's face and body from showing the signs of growing older.[21]

Many American women in the late 1960s and early 1970s still felt bound by traditional social roles and cultural expectations. While women both young and old had begun to speak out against women's oppression, there were many who remained ignorant of or opposed to the possibility of major changes in the social, cultural, and political situation of women in America. Surely, estrogen helped many women to cope with the unpleasant physical symptoms of menopause, such as hot flashes, night sweats, and painful intercourse. And those who claimed that estrogen helped them to "feel" and "look" young (or younger than their nonestrogenated selves) sincerely believed in its restorative powers. For such women, an important objective was to maintain their youthful appearance and demeanor so as to remain attractive and pleasing to men, and they trusted their (largely male) physicians to provide them with a medical regimen that would help them to meet that goal.

Physicians' inclination to prescribe estrogen was further encouraged by the aggressive marketing techniques of pharmaceutical manufacturers. Since these companies were not permitted to advertise prescription drugs directly to the public, they undertook a variety of efforts to make their estrogen products known to doctors. Starting in the 1930s, they supplied the drugs free of charge to researchers, who then acknowledged the companies' generosity in the resulting articles published in medical journals. Manufacturers also advertised in the pages of medical journals, sent product information by direct mail to physicians, and employed special salesmen—called detailmen—to visit physicians and deliver the information in person. By the late 1960s, Ayerst Laboratories was spending a million dollars a year to advertise Premarin.[22] These efforts helped Premarin, the most popular brand of estrogen, to become one of the top five prescription medications in the United States by 1975.[23] The annual number of estrogen prescriptions increased from 15.5 million in 1966 to 28 million in 1975, when estrogen sales hit $85 million in that year alone.[24]

It is important to recognize, as historian Judith Houck has pointed out, that women did not necessarily believe that estrogen was a miracle drug; many of those who took the hormones may have done so purely as a practical measure to get some relief from unpleasant symptoms such as hot flashes and genital atrophy.[25] The point here is that women had to make health care decisions based on what their doctors chose to tell them and what they gleaned from scattered reports in the popular press. However, they also had to make these decisions in the context of a barrage of dominant cultural representations of beauty and persuasive cultural messages about the status of aging women in American society.[26]

That women recognized the paucity of balanced information available to them on the subjects of menopause, hormones, and aging was revealed in the responses to a survey questionnaire sent out to 500 women by the Boston Women's Health Book Collective. One question asked, "In what ways do you feel that your own education about menopause has been good or bad?" While a few women considered themselves to be well-informed, many more commented on the dearth of available information. Respondents described their education as "very bad," "too sparse," "negligible," "inadequate," "non-existent and difficult to get."[27] One women replied angrily, "What "education"? We don't get any on the subject, and doctors are the *last* to admit you're going through menopause or are starting it!"[28] Another echoed the sentiment that physicians withheld information: "It has been scanty because doctors don't always tell and there isn't enough material published."[29] Several described themselves as self-educated, reading up on the subject in whatever resources they could find: "I read in the newspapers and magazines, and once sent away for a booklet."[30] These responses reflected the limited availability of educational materials on the symptoms, sequelae, and treatment of menopause. In the 1960s and early 1970s, the provision of health information fell to physicians and pharmaceutical manufacturers. Journalists took their cues from medical experts, and government regulators remained silent on the subjects of menopause and hormone replacement therapy.

Changing Views on Estrogen: Medical Reports and the Women's Health Movement

This situation changed after December 1975, when two studies, published in the same issue of the *New England Journal of Medicine*, offered conclusive evidence that estrogen users were more likely to get endometrial cancer as compared to nonusers.[31] Furthermore, it appeared that the longer a woman took estrogen, the greater her risk of endometrial cancer (up to 14 times for women who took the hormones for more than seven years).[32] These results were corroborated by two more studies published in the same journal the following June. None of these studies offered definitive proof of a causal relationship between estrogens and endometrial cancer. At best, they provided evidence of an association between the use of estrogen and the development of endometrial cancer. In spite of this ambiguity, the statistics presented in these reports cast a pall on both the short-term and long-term prescription of estrogen replacement therapy. Within five years, the number of estrogen prescriptions written annually dropped by 50 percent, to 14 million.[33] Thanks to extensive media coverage of the estrogen-cancer link, women became apprehensive about hormone therapy. Physicians, too, became more cautious with regard to the use of estrogen; one survey found that the majority still recommended estrogen to relieve the symptoms of menopause (such as frequent, severe hot flashes), but they tended to prescribe smaller doses for shorter periods of time. Few advocated the use of high-dose estrogen for more than six months or low-dose estrogen for more than three years.[34]

In 1976, the U.S. Food and Drug Administration (FDA) mandated a patient labeling requirement for estrogen.[35] The goal of this FDA action was to provide patients with enough information to help them participate in the decision whether or not to use long-term hormone therapy. Manufacturers were compelled to supply, and pharmacists to distribute, a lengthy leaflet that discussed the uses, dangers, and side effects of estrogen. The pamphlet warned the patient in explicit detail about the increased risk of endometrial cancer, as well as the potential risk of breast cancer, gall bladder disease, and abnormal blood clotting.[36]

As they reported these scientific findings and government proceedings, magazine writers became more cautious in their treatment of menopause-related topics. They also recognized the importance of self-education in health matters . . . one of the central tenets of the women's health movement. Whereas the readers of *Harper's Bazaar*'s 1973 "Over-40 Guide on Health, Looks, and Sex" were told: "It is up to your doctor to decide . . . whether you should have estrogen and if so how much, how often, for how long," articles just a few years later encouraged women to become active participants in the decision whether or not to treat menopause with hormones. A 1977 article in *McCall's*, titled "Estrogen: The Rewards and Risks," asked, "On what basis can a woman make her decision about the use of estrogen therapy if she cannot rely upon her physician's knowledge?" It went on to reply, "Fortunately, for a woman to educate herself about menopause and the use of estrogen replacement therapy is not all that difficult."[37] In 1973, the balance of power in the doctor-patient relationship lay with the physician; by the end of the decade, female patients, armed with information, had begun to tip the scales, certainly not entirely in their favor, but at least toward a more equitable distribution of influence.

To educate themselves about menopause and hormones, many women turned to two popular books published in 1977: *Menopause: A Positive Approach*, by Rosetta Reitz, and *Women and the Crisis in Sex Hormones*, by Barbara Seaman and Gideon Seaman.[38] Both books were hailed by the women's health movement for their honesty and comprehensiveness, and both went through several paperback reprintings. *Women and the Crisis in Sex Hormones* was a featured alternate of the Book-of-the-Month Club and was serialized in *Ms., Family Circle, Working Woman,* and *Playgirl.* In its first six months, the book sold 60,000 copies; the Erie (PA) *Times-News* reported that it was the second "best-read" nonfiction book in Erie County, based on library requests.[39]

Barbara Seaman had made a name for herself several years earlier, when the publication of her first book, *The Doctors' Case Against the Pill*, instigated Senate hearings in 1970 on the safety of oral contraceptives.[40] As one of the founders of the National Women's Health Network, Seaman fought on the front lines of the movement to increase women's access to information about health care issues. Reitz, on the other hand, came to research menopause out of frustration, because her doctor was "too busy" to answer her questions and the few books she was able to find on menopause portrayed it negatively and simplistically, as resulting either from physiological hormone deficiency or the social stresses of being middle-aged. Her book evolved out of her personal quest for information, which she found in a series of "menopause workshops." Reitz gathered midlife women at her home for weekly consciousness-raising sessions; those intimate discussions—about menopause, aging, love, sex, health, work, children—provided Reitz with the information she wanted to share with a broader audience.[41]

Both Reitz and the Seamans rejected the medicalization of menopause, particularly the use of estrogen replacement therapy. Not only did these authors present their readers with alternative approaches to dealing with symptoms of menopause, but they also offered an alternative source of information, more comprehensive than a magazine article, a promotional brochure, or a chat in the doctor's office. These books struck a responsive chord with women. After Barbara and Gideon Seaman appeared on the "Phil Donahue Show" in September 1977, dozens of women wrote to ask more questions, to request a copy of the book, or to express their appreciation. One viewer from Pennsylvania wrote, "I saw you on Donahue this A.M. You're a refreshing breath of fresh air in our pill polluted society."[42]

Reitz and the Seamans represented one end of the spectrum of opinion on the management of menopause. Physicians wrote their own books to encourage women to seek the counsel of medical specialists and perhaps to regain their authority in what they considered to be their purview. An advertisement for *No Pause at All,* by Louis Parish, M.D. (authors with medical degrees were always clearly identified as such), in the *New York Times* announced, "a renowned endocrinologist/psychiatrist shows you that there need be no pause at all in the normal functioning of your physical and psychological being. He explains all the tremendous strides medicine has made in the cure and prevention of middle age discomfort."[43] *The Menopause Book* promised "up-to-date helpful medical facts by eight women doctors"; advertising copy for this volume made much of the gender and expertise of its authors.[44] These books sought to reassure women that physicians and pharmaceuticals should continue to play a central role in guiding women through menopause. Celebrities also weighed in with their own advice manuals; their authority in this area presumably rested on their ability to maintain their good looks. In her book, *Winning the Age Game,* model and television personality Gloria Heidi "share[d] the secrets of her own Ageless Woman look" in "her new, step-by-step, comprehensive guide."[45] In the late 1970s, women scanning the bookshelves to read up on menopause and mid-life health issues found a greater selection than in previous decades.

The provision of menopause information soon expanded beyond printed sources, as the notion that women should seek out health education began to take hold. In the early 1980s, local hospitals and Planned Parenthood affiliates offered menopause seminars; some women started their own menopause discussion groups to "counter the negative attitudes of doctors."[46] These support groups sprang up around the country, modeled, like Reitz's workshops, on the women's liberation consciousness-raising rap groups of the previous decade. In the absence of a definitive medical consensus, women turned to each other to share their experiences of menopause and aging.

Older members of the Boston Women's Health Book Collective also recognized that *Our Bodies, Ourselves* did not adequately address aging. As Paula Brown Doress and Diana Laskin Siegal, collaborators on a new chapter called "Women Growing Older," recalled, "the task of squeezing the health and living issues of four or five decades of life into one chapter seemed insurmountable." They realized that the second half of life deserved a book of its own, which led to the publication in 1987 of *Ourselves, Growing Older: Women Aging with Knowledge and Power.*[47] The book was predicated on the feminist principle that knowledge equals power. The better educated women were about menopause and aging, the authors reasoned, the more able they would be to take charge of their lives. In the foreword to the volume, Tish Sommers, one of the founders of OWL, the Older Women's League, wrote:

> Knowledge is the first step. The second is taking more control over whatever aspects of our health that we can. From dependency on "experts" we must move toward greater independence, which means taking on a greater degree of responsibility for our own bodies, as well as for the policies which affect us.[48]

The authors of *Ourselves, Growing Older* anticipated that the generation of women who had learned about their bodies from *Our Bodies, Ourselves* would turn to this sequel as they grew older.

Other publications arose in the 1980s to deal specifically with health-related issues. These feminist health newsletters, with names like *Hot Flash* and *A Friend Indeed for women in the prime of life,* were also conceived as educational resources for midlife and older women. However, the impact of the messages conveyed in feminist newsletters was limited because they reached a relatively small number of women. The newsletters—started in the days before desktop publishing really took off—had at most a few thousand subscribers. *Ourselves, Growing Older* fared better. The first edition sold 100,000 copies (from 1987 until 1994) and the second edition (from its publication in 1994 through 2004) sold 87,000.[49] It was especially difficult for alternative models to compete with the medicalization of menopause and aging and the widespread dissemination of this model in the mainstream popular media, whose audiences were measured in the millions.

The information delivered by the mass media did not fully reflect the changes occurring within the women's movement regarding age activism and health activism. After the flurry of interest in menopause and estrogen replacement therapy in the wake of the endometrial cancer scare had subsided, magazines and newspapers once again devoted little space to these topics. Psychologists Linda Gannon and Jill Stevens found that in the 1980s twice as many articles were published on infertility, which affects a relatively small proportion of women, as were published on menopause, which eventually affects all women. Furthermore, those articles that did appear in popular periodicals continued to address menopause as a medical condition in need of pharmaceutical therapy.[50]

HRT Makes a Comeback

The medical model of menopause and aging gained further support in the 1980s as the preventive effect of estrogen on osteoporosis received more attention and endorsement. As the *New Republic* cynically observed, osteoporosis became "the disease of the week."[51] In 1982, Ayerst Laboratories hired the public relations firm of Burson-Marsteller to educate the public about osteoporosis via radio, television, and magazine articles.[52] The firm also produced pamphlets and seminar materials for health professionals to disseminate to women. Of course, these "educational materials" also discussed the use of estrogen (in the form of Ayerst's Premarin) to combat the ravages of this disease. In 1984, the National Institutes of Health convened a consensus group to develop a statement on osteoporosis. This group asserted that estrogen replacement was the most effective way to prevent bone loss in women and advised physicians to consider hormone replacement for all women so long as they had no contraindications, understood the risks, and agreed to regular medical examinations.[53] Two years later, the FDA formally announced the conclusion that estrogen was considered effective for treating postmenopausal osteoporosis. By redirecting the attention of both doctors and women to the relationship between estrogen loss and bone loss, manufacturers and estrogen advocates successfully revived the long-term market for hormone therapy. The endometrial cancer scare had circumscribed the use of hormones to the short-term relief of unpleasant menopausal symptoms. The new epidemiological evidence about osteoporosis (and the finding that the addition of progestin to estrogen in the hormone replacement regimen prevented the development of endometrial cancer) justified the reexpansion of clinical indications for postmenopausal hormone therapy to include the long-term prevention of debilitating disease.

Magazine writers, provided with abundant information by Ayerst's publicity campaign, were quick to popularize the link between osteoporosis and the decreased production of estrogen after menopause. Whereas articles in the late 1970s had expressed caution and concern about hormone replacement therapy, those written a decade later touted its benefits for aging women. Articles about estrogen in news weeklies and women's magazines echoed the optimism of the 1960s. According to *U.S. News & World Report*:

> Many women secretly fear that menopause is the beginning of the end of the good life: Hard on the heels of the hot flashes and mood shifts follow the frailties and indignities of old age. Now, experts say that doesn't have to be so. Regular doses of estrogen and other hormones—long regarded as a last-ditch only treatment for menopause's discomforts—may actually launch a woman into vigorous senior citizenship by strengthening her bones, heart, and libido.[54]

Estrogen received an increasingly positive presentation in several popular magazines as the 1980s progressed. Women's magazines such as *Redbook* and *Ladies' Home Journal* reached 14 million and 19 million readers each month, respectively, while 20 million people read *Newsweek* and 30 million leafed through the pages of *People* every week.[55] Articles that touted the osteoporosis-preventing and anti-aging properties of estrogen were far more widely distributed—and thus had a far greater influence—than anything written in *Ourselves, Growing Older* or other feminist publications.

Pharmaceutical companies subtly promoted the medical management of menopause and the use of hormone replacement therapy in pamphlets produced in the 1980s, while paying lip service to the value of self-education for women. A booklet put out by Mead Johnson Laboratories, entitled, "For the Woman Approaching Menopause," advised:

> Menopausal women should keep up to date with the latest thinking about menopause from reputable professional sources. Avoid forming your judgments concerning management of your menopause on the basis of casual chats with friends, hearsay, and sensationalist media reports.[56]

The message here was that there were both good and bad sources of information; medical experts were "right," nonprofessionals were "wrong," and women should steer clear of the latter. It was acceptable for women to pursue their own education in health matters, so long as they sought information from within the medical-pharmaceutical community.

Nonetheless, women still looked to the media and to each other for information. In December 1986, as part of an article on the new estrogen skin patch, the *Saturday Evening Post* included a 13-question survey about menopausal and postmenopausal symptoms, osteoporosis, and the use of estrogen.[57] Within two weeks, 1,822 women had returned their completed surveys. The *Post* tabulated these results and published them in the next issue of the magazine, as part of an article entitled, "More About Estrogen Skin Patches," along with interviews with women who switched from estrogen pills to the patch and advice from Dr. Lila Nachtigall, who had directed a clinical trial of the transdermal patch.[58] Clearly biased in favor of estrogen and the medical management of menopause and post-menopause, this article typified popular magazine coverage of the subject in the mid-1980s. The article did note that Nachtigall advised women to try nonprescription remedies before opting for hormone therapy, but it also reprinted six paragraphs verbatim from her book, *Estrogen: The Facts Can Change Your Life*, on the need for women over 40 to consult regularly with physicians, general practitioners, gynecologists, and, if necessary, reproductive endocrinologists. Menopause, according to Nachtigall, ought to be medically supervised, and the physician-editor of the *Saturday Evening Post* concurred.

The results of the survey supported the success of the current medical model; of course, the respondents were a self-selected group and not at all representative of the larger population of midlife and older women in America. Why would a woman bother to fill out and mail in the survey? Altruism, perhaps, since the survey was introduced with the notice that "Your willingness to participate will help discover better ways of treating the very real problems that some women suffer during and after menopause."[59] Since most of the respondents simply checked the yes or no spaces on the forms, it is hard to know their motivation for participation. The results indicated that many of them had experienced discomfort during or after menopause. Of the 1,822 women, 78 percent reported experience with hot flashes, and more than half said they suffered anxiety (55%), irritability (55%), or depression (51%). Two-thirds had taken estrogen at one time or another for menopausal or postmenopausal reasons. Although only 7 percent reported a vertebral fracture and 2 percent a broken hip, 29 percent claimed to have lost height and almost half (45%) planned to take long-term estrogen replacement therapy to prevent osteoporosis. Interestingly, the original survey was called "Menopausal and Postmenopausal Women's Survey," but the report the following month was titled, "Preliminary Results of *Estrogen* Study" [emphasis added]. The one-page summary listed the raw data: the number who replied yes or no or who gave no answer to each of the questions. Although the heading of "preliminary" hinted that a more complete final report might follow, no such analysis ever appeared in the *Post*.

However, the responses kept flooding in. Altogether, the *Post* recorded the receipt of 4,570 replies.[60] In addition to checking the yes or no spaces on the survey form, many women provided additional information by elaborating in the margins or on extra sheets of paper, as the form allotted no extra space for comments. These unsolicited annotations revealed a wide spectrum of attitudes and practices concerning hormone replacement therapy. Even within the relatively narrow demographic slice of America represented by the *Saturday Evening Post* readers who took the time to fill out and return the survey—that is, mainly middle-class, middle-aged white women—there was a great range of positions not only on hormone use, but also on the roles of doctors and patients in the management of menopause and postmenopause, the importance of the dissemination of health information for lay audiences, and conceptions of female aging.[61] The fact that thousands of women chose to complete the surveys demonstrated that women were interested in sharing their experiences of menopause and aging and learning about those of others; the added comments on hundreds of these forms further uncovered the influences on women's interpretations of their experiences coming from the contemporary proestrogen climate in both the medical community and the popular press and from competing feminist perspectives.

Above all, the 1986 *Saturday Evening Post* survey revealed that the menopausal and postmenopausal experiences of American women—even when taken from a relatively homogeneous population subgroup—could not be reduced to a few simple archetypes. Some were informed by the medical model and some by the feminist model; some incorporated parts of each paradigm into their health care decision making, and some were guided wholly by their own principles. Women interpreted menopause as one thread within the fabric of their lives, intertwined with other experiences, such as getting divorced or going back to school or taking on a new job or caring for elderly parents or dealing with financial problems. What is clear, however, is that a majority (of this sample, at least) did turn to medicine for relief from physical discomfort or psychological distress during and after menopause. And more often than not, the treatment prescribed by physicians was hormone replacement. In spite of the so-called cancer scare of the previous decade, many women seemed to acknowledge the potential short-term and long-term benefits of estrogen, thanks, in part, to upbeat articles in mainstream magazines that brought news from the world of medical science of the rehabilitation of estrogen.

The Baby Boomers Reach Menopause

In the next decade, menopause came "out of the closet." Although it had received attention in the mainstream media in the past, there was an explosion in the quantity of information that circulated in the 1990s, as the women of the baby boom generation began to approach the age of menopause. Because of their sheer force of numbers, baby boomers had been able to focus public attention on their shifting interests and concerns throughout the stages of their lives. This largest generation has made its influence felt in every decade: as teens in the 1960s, as young adults in the 1970s, as "yuppies" in the 1980s, and as they entered middle age in the 1990s. And just as birth control and abortion topped the women's health agenda in the 1970s, menopause became a priority two decades later, as women demanded information about this next chapter in their lives.

Women who sought to educate themselves often had to look no further than the local bookstore or newsstand. A spate of mass-market books appeared, led by Gail Sheehy's *Menopause: The Silent Passage* in 1992.[62] Based on interviews with both women and physicians as well as her own personal experience, Sheehy attempted to break "the conspiracy of silence" about menopause. She described the various conditions attributed to menopause (from hot flashes and vaginal atrophy to heart disease and osteoporosis) and then debated hormone replacement therapy in a chapter called "Should I or Shouldn't I?" Sheehy presented the pros and cons of long-term hormone use (and alternative therapies) and left it up to her readers to decide for themselves. Clearly, there was a market for more information about menopause: Sheehy's book was the ninth best-selling nonfiction book of 1992; it spent a total of 114 weeks on the *New York Times Book Review*'s list of best sellers.[63]

By 1995, more than a hundred books had been published by both medical and lay authors on the subject of menopause.[64] Newspaper and magazine editors sensed the sea change in interest about menopause and began to publish more articles on menopause and related subjects (such as hormone replacement therapy and osteoporosis). In all of the popular magazines indexed in the *Reader's Guide to Periodical Literature,* there was never more than a total of a dozen articles on menopause or related topics in any year prior to 1985. The numbers began to rise in the late 1980s, and by the 1990s, there was an average of more than 32 articles a year. Articles with titles such as "The Baby Boom Meets Menopause" and "Menopause and the Working Boomer" recognized that "an entire generation is approaching a big change."[65]

These articles often addressed the controversy over hormone replacement therapy, quoting the latest statistics and "expert" opinions on the relationships between estrogen and endometrial cancer, breast cancer, osteoporosis, and heart disease. In 1991, the National Women's Health Network surveyed a sample of 25 articles from popular magazines and found that most (83%) gave a balanced assessment, mentioning both risks and benefits of hormone replacement therapy, and more than half (52%) suggested alternative therapies, such as herbal remedies, dietary options, and exercise regimens.[66] Some articles offered handy quizzes for women to assess for themselves the relative risks and benefits of estrogen replacement therapy. Although *McCall's* included with its "estrogen test" the caveat that "this is, of course, only a starting point for further discussion . . . with your physician," the implication was that women could and should learn enough about menopause, aging, and hormones to engage in intelligent conversations with their doctors.[67] *Better Homes and Gardens* concurred:

> Educate Yourself. In the absence of complete medical data, the best advice is to learn all you can about hormone replacement and make a careful assessment of your own risks . . . the biggest health risk of all—not being an educated consumer.[68]

The language of this recommendation is significant in its address of women as active consumers in charge of their health care, not as passive patients at the mercy of their doctors.

Television and radio jumped on the menopause bandwagon, too. Whereas the subject of menopause was rarely heard on the airwaves prior to 1990, it rapidly became a popular topic on news broadcasts and talk shows as the decade progressed.[69] Paula Zahn, cohost of "CBS This Morning," introduced her 1990 interview with the authors of a book called *Managing Your Menopause* by saying, "Women have traditionally dreaded approaching menopause. Now medical science and a woman's own awareness can eliminate much of that dread and many of the problems."[70] A 1992 CBS Evening News "Eye on America" segment explored how baby boomers on the verge of menopause were "demanding explanations from doctors and talking openly among themselves."[71] Even situation comedies broached the subject, with menopause featured in episodes of "The Golden Girls" and "The Cosby Show."[72] And, with restrictions lifted on direct-to-consumer advertising of prescription drugs on television in 1997, Wyeth began to run commercials for Premarin during prime time. In the early 2000s, ex-supermodel Lauren Hutton starred in one television spot for Premarin, in which she proclaimed, "Knowledge is power. Information is how you get it."[73] She advised women to ask their doctors for that information and offered her own experience as a cautionary tale: having lost one inch in height in the year after menopause, she decided to take her doctor's advice and started a regimen of HRT.

This expanded coverage meant that more women were exposed more often to talk—in news, in entertainment, in advertising—about menopause and hormone replacement therapy. Learning now included a sort of passive reception: while watching the evening news or listening to a daytime talk show, a woman might hear something about menopause—a brief news report, an in-depth story, or perhaps an interview with an author of the latest self-help guide. The point here is that the recipient of this information was not actively seeking knowledge; as menopause became popularized, women had to work less hard to learn about it.

Of course, for the woman who wanted to conduct her own research, the available information had multiplied dramatically since the previous decade. Not only were there more books and articles available in print, from both mainstream and alternative presses, but also the new medium of the internet provided a whole new educational resource. For example, the web site called "Power Surge" came on line in 1994, offering women the opportunity to chat via email with doctors, authors, and other women going through menopause. The heading to one of the links provided by this site proclaims, "Knowledge is Power. Educate your body by educating yourself."[74]

While many observers welcomed this influx of information, the feminist magazine, *Ms.,* scoffed at the increased attention to menopause. A 1993 article began:

> 1992 was the year of the . . . full-scale launch of the meno-boom, inundating women with "information" [the quotation marks signifying irony] that detailed all the loss, misery, humiliation, and despair suddenly in store for us.[75]

And a feminist columnist for *The Nation,* a liberal weekly publication, sat-
irized the plethora of magazine and newspaper articles as follows:

> Invariably beginning with a tribute to upper-income baby-boom women—
> you know, educated medical consumers used to speaking out and getting
> what they want—the article moves in swiftly for the kill: hot flashes, osteo-
> porosis, irritability, heart disease, insomnia, migraines, declining libido,
> wrinkles and job loss.[76]

Why did the latest batch of menopause articles merit such derision from
these commentators?

Journalists writing in the 1980s and 1990s has successfully incorporated
the rhetoric of the women's health movement, encouraging their readers
to become educated health consumers. However, many articles still per-
petuated the idea of medicalized menopause and portrayed the image of
menopausal women "in decline," if not in the main text itself, then in the
accompanying illustrations. *Newsweek*'s cover story on menopause in 1992
graced its cover with the image of a woman's face represented as a tree los-
ing its leaves (figure 1); three years later, *Time*'s cover story on "The Estrogen Dilemma" included a similar drawing (figure 2).[77] The autumnal
imagery of deciduous tree branches (accompanied by the troubled look
on the woman's face) presented menopause and the years beyond as cause
for concern. This tendency to biologize menopause and to suggest personal health remedies (pharmaceutical, nutritional, or otherwise) rather
than to evaluate the social status of older women irritated feminists who
had hoped for broader cultural change.

Indeed, many health professionals sought to use educational materials
to encourage more older women to use hormone replacement therapy; in
their opinion, an informed patient was one who agreed that the benefits of
estrogen outweighed the risks. The authors of one study of the use of hormone replacement therapy in the mid-1990s concluded, "Perhaps the
most important goal, then, is to educate menopausal women to the facts
as we now know them—that the risk of cancer appears to be quite small
relative to the benefits of HRT [hormone replacement therapy] in preventing CVD [cardiovascular disease], the number one cause of death of
women in America."[78] Another survey cautioned that "unless physicians
and other women's health providers seek to meet these needs [for information about the risks and benefits of HRT], this important educational
task may be left to less-informed sources."[79] Physicians worried that alter-

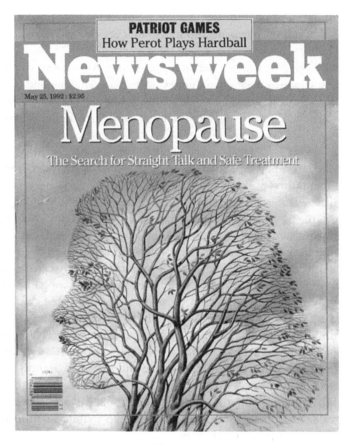

Figure 1.

native sources would turn women away from the medical model of menopause and aging.

The author of the *Ms.* magazine article declared, "Our main goal should be to take control of the flood of "news" about menopause,"[80] but such manipulation by just one interest group (even one so large as the community of older women) was not possible when so many other factors were involved. Second wave feminism and the aging of the baby boom generation were not the only forces to influence media coverage of information about menopause and aging. The clout of the powerful drug industry, the enduring authority of the medical profession, and the pres-

sure of the prevailing youth-oriented culture in America helped to construct both the content and the presentation of health information for
older women. Informal education about menopause and aging, as
imparted in the pages of popular periodicals or in television interviews,
for example, entailed much more than the simple communication of fac-
tual information. Rather, these educational sources presented a myriad of
contested meanings and conflicting messages about the health and social
aspects of growing older as a woman in America.

Although women today have access to much more health-related information than any of their predecessors, these resources do not seem to clear

Figure 2.

the path through menopause and postmenopause.[81] Three factors have contributed to the lack of consensus about menopause and its management among both the producers and consumers of information on these topics. First, the presentation and interpretation of advice differed according to the author's (and reader's) position on the medicalization of menopause and postmenopause. While menopause and its sequelae, like other aspects of women's health (e.g., childbirth, birth control), increasingly came under medical control in the 1900s, there was also a parallel trend in recent years to redefine menopause as a "natural" event. Pharmaceutical manufacturers, not surprisingly, encouraged the readers of their educational brochures, advertisements, and other promotional materials to consult their doctors about hormone replacement therapy. Feminist health writers, on the other hand, argued that aging was a natural process, which a woman could "manage" herself, with diet, exercise, and spiritual, emotional, and intellectual fulfillment. Popular magazine articles often combined these two approaches, offering information about both medical and nonmedical regimens, leaving the reader to make her own decisions.[82]

Second, the decision whether or not to use hormone replacement therapy was further complicated by the fact that estrogen was one of a class of preventive drugs with ambiguous results. Neither patients nor physicians nor researchers nor government regulators had enough information to ascertain clearly the benefits and risks of long-term hormone replacement therapy. This is not to deny the therapeutic value of estrogen in relieving the short-term symptoms of menopause, such as hot flashes and night sweats. However, the decision to take hormones for 10 or 20 or 30 years after menopause was more difficult, as epidemiological statistics did not necessarily translate into clear choices for individuals.

Finally, the personal and cultural implications of the medicalization of aging and the dilemma of growing old in a youth-centered society presented issues that remain unsettled and unsettling. Many of the available educational resources on menopause and aging gave equal weight to issues of health and appearance. Women in postmenopause were threatened not only with osteoporosis and heart disease, but also with wrinkled skin and sagging breasts. The dominant cultural valuation of youth as the standard for beauty encouraged women to try to impede the aging process, not only to prevent disease but also to forestall the physical signs of growing old.[83] Ambivalence about the status of older women in American society further confounded the messages implicit within allegedly neutral health information sources.

By the end of the twentieth century, the feminist mandate for more information about the topics of menopause and hormone replacement therapy to be made freely available in the public domain had come to pass, but the substance of those reports did not incorporate a very wide spectrum of possible approaches to health concerns in the years during and after menopause. Although journalists acknowledged the existence of nonhormonal therapies and even the possibility of no therapy at all, the medical-pharmaceutical paradigm of treating midlife and older women with hormones remained dominant, at least until the conclusion of the Women's Health Initiative in 2002. But ambivalence persisted within the medical community, and these uncertainties were passed on to consumers via the media. The production and consumption of health information on menopause and HRT was complicated by the multiple medical and popular interpretations of the available data, as well as the various contexts in which those data were translated for the general public. Unfortunately, in this case, more information and greater access to that information did not necessarily produce a wiser, better informed population.

NOTES

The author would like thank Andrea Tone for her helpful suggestions in the preparation of this chapter and to acknowledge fellowship support from the National Academy of Education and the Spencer Foundation.

1. Prescription data come from the annual list of top 200 prescription drugs, published in *American Druggist* until that journal ceased publication in 1999, and the annual list of top 200 drugs published on the internet website RxList (http://www.rxlist.com/top200.htm) from 1999 to 2004. The figures on the decrease in Premarin sales come from NDCHealth's website (http://www.ndchealth.com/index.asp). See also Adam L. Hersh, Marcia L. Stefanick, and Randall S. Stafford, "National Use of Postmenopausal Hormone Therapy: Annual Trends and Response to Recent Evidence," *Journal of the American Medical Association* 291 (7 January 2004): 47–53.

2. See, for example, Rita Rubin, "U.S. Halts Study on Hormone Therapy," *USA Today* (9 July 2002): 1A; Gina Kolata, "Study Is Halted Over Rise Seen In Cancer Risk," *New York Times* (9 July 2002): 1; Patricia Guthrie, "Study says halt hormone therapy," *Pittsburgh Post-Gazette* (9 July 2002): 1; Cheryl Clark, "Risks cited in hormone therapy for women; Harm reportedly greater than benefit; Trial halted" *San Diego Union-Tribune* (9 July 2002): A-1.

3. Elizabeth Siegel Watkins, "Dispensing with Aging: Changing Rationales for Long-term Hormone Replacement Therapy, 1960–2000" *Pharmacy in History* 43, no. 1 (2001): 23–37.

4. Diane E. Wysowski, Linda Golden, and Laurie Burke, "Use of Menopausal Estrogens and Medroxyprogesterone in the United States, 1982–1992" *Obstetrics & Gynecology* 85 (1995): 9.

5. Yankelovich, Skelly & White, "General Mills American Family Study 1978–79: Family Health in an Era of Stress" (October 1978). Roper Center for Public Opinion Research, University of Connecticut, Storrs, CT.

6. Phyllis Kernoff Mansfield and Ann M. Voda, "From Edith Bunker to the 6:00 News: How and What Midlife Women Learn about Menopause" *Women & Therapy* 14 (1993): 97–98.

7. Ibid.: 98–99.

8. Julia Kagan and Jo David, "The Facts of Life: What Every Woman Over 35 Needs to Know about Her Body" *McCall's* 118 (June 1991): 71.

9. The *New York Times* reported in late 1975 that Premarin accounted for 75–80 percent of the estrogen market. See Robert Metz, "Market Place" *New York Times* (12 November 1975): 64. Two years later, *Chemical Week* reported that Premarin held 63 percent. See "Estrogens hurting, corticoids healthy" *Chemical Week* (23 November 1977): 23. A 1985 study from the Drug Use Analysis Branch of the Food and Drug Administration estimated that Premarin had accounted for 70 percent of estrogen use over time. See Dianne L. Kennedy et al., "Noncontraceptive Estrogens and Progestins: Use Patterns Over Time," *Obstetrics and Gynecology* 65 (March 1985): 444. In testimony at a 1976 Senate hearing investigating the association between estrogen and endometrial cancer, an Ayerst executive claimed that Premarin held less than 40 percent of the estrogen market share. Chester J. Cavallito, Executive Vice-President for Scientific Affairs, Ayerst Laboratories, quoted in U.S. Senate, 94th Congress, 2nd Session, *Oral Contraceptives and Estrogens for Postmenopausal Use,* Joint Hearing before the Subcommittee on Health of the Committee on Labor and Public Welfare and the Subcommittee on Administrative Practice and Procedure of the Committee on the Judiciary (January 21, 1976), 213.

10. Judith A. Houck, "Common Experiences and Changing Meanings: Women, Medicine, and Menopause in the United States, 1890–1980," Ph.D. dissertation, University of Wisconsin-Madison, 1998, 165.

11. Ibid., 161–165.

12. Sherwin A. Kaufman, M.D., "The Truth about Female Hormones" *Ladies' Home Journal* 82 (January 1965): 22.

13. Ann Walsh, "Pills to Keep Women Young" *McCall's* 93 (October 1965): 104.

14. "Bazaar's Over-40 Guide on Health, Looks, Sex" *Harper's Bazaar* 106 (August 1973): 87.

15. Houck, "Common Experiences and Changing Meanings," 306.

16. Grace Naismith, "Common Sense and the 'Femininity Pill'" *Reader's Digest* 89 (September 1966): 99.

17. "Estrogens During and After the Menopause" *Medical Letter* 7 (July 1965): 56.

18. Julia Kagan, ""Hormone Therapy at Menopause: What Women Doctors Prescribe and Take" *McCall's* 103 (October 1975): 33.

19. Ibid.: 34.

20. Lindsay R. Curtis, M.D., "The Menopause: a new life of confidence and contentment" (Bristol, Tennessee: SeMed Pharmaceuticals, 1969): 18–19.

21. Susan Sontag, "The Double Standard of Aging," *Saturday Review,* 23 September 1972: 31, 35.

22. "Ayerst Premarin Goes to Sudler & Hennessey" *New York Times* (6 January 1970): 64.

23. Patricia A. Kaufert and Sonja M. McKinlay, "Estrogen-Replacement Therapy: The Production of Medical Knowledge and the Emergence of Policy," in Ellen Lewin and Virginia Olesen (editors), *Women, Health, and Healing: Toward a New Perspective* (New York: Tavistock Publications, 1985), 114.

24. Kennedy et al., "Noncontraceptive Estrogens and Progestins," 442.

25. Houck, "Common Experiences and Changing Meanings," 330.

26. For a discussion of the power of cultural imagery and messages in the construction of the body, see Susan Bordo, *Unbearable Weight: Feminism, Western Culture, and the Body* (Berkeley: University of California Press, 1993).

27. The questionnaire and 146 responses can be found in the Records of the Boston Women's Health Book Collective at the Schlesinger Library, Cambridge, Massachusetts (hereafter, Records-BWHBC), Accession #99-M147, carton 8. #73, #49, #444, #401, #455, Records-BWHBC, Accession #99-M147, carton 8.

28. #459, Records-BWHBC, Accession #99-M147, carton 8.

29. #400, Records-BWHBC, Accession #99-M147, carton 8.

30. #433, Records-BWHBC, Accession #99-M147, carton 8.

31. Donald C. Smith et al., "Association of Exogenous Estrogen and Endometrial Carcinoma," *New England Journal of Medicine* 293 (4 December 1975): 1164–1167; Harry K. Ziel and William D. Finkle, "Increased Risk of Endometrial Carcinoma Among Users of Conjugated Estrogens," *New England Journal of Medicine* 293 (4 December 1975): 1167–1170.

32. Thomas M. Mack et al., "Estrogens and Endometrial Cancer in a Retirement Community," *New England Journal of Medicine* 294 (3 June 1976): 1262–1267; Noel S. Weiss et al., "Increasing Incidence of Endometrial Cancer in the United States," *New England Journal of Medicine* 294 (3 June 1976): 1259–1262.

33. The number of estrogen prescriptions dropped from 28 million in 1975 to 14 million in 1980. Kennedy et al., "Noncontraceptive Estrogens and Progestins," 443.

34. Beverly H. Pasley et al., "Prescribing Estrogen During Menopause: Physician Survey of Practices in 1974 and 1981," *Public Health Reports* 99 (July-August 1984): 424–429.

35. *Federal Register*, 41, 29 September 1976: 43108–43109. For more on the estrogen patient package insert, see Elizabeth Siegel Watkins, "'Doctor, are you trying to kill me?': Ambivalence about the Patient Package Insert for Estrogen" *Bulletin of the History of Medicine* 76 (Spring 2002): 84–104.

36. *Federal Register*, 42, 1977: 37642; *Federal Register*, 43, 1978: 4223.

37. Paula Weideger, "Estrogen: The Rewards and Risks" *McCall's* 104 (March 1977): 70.

38. Rosetta Reitz, *Menopause: A Positive Approach* (New York: Chilton Book Company, 1977); Barbara Seaman and Gideon Seaman, M.D., *Women and the Crisis in Sex Hormones* (New York: Rawson, 1977). Seaman's book was the third in what she called her "estrogen trilogy," following *The Doctors' Case Against the Pill* (1969) and *Free and Female* (1972). After writing a biography of the author Jacqueline Susann (*Lovely Me*, 1987), Seaman returned to the topic of estrogen in her 2003 book, *The Greatest Experiment Ever Performed on Women: Exploding the Estrogen Myth* (New York: Hyperion, 2003).

39. Sales figures are from *Playbill* (March 1978): n.p.; "Best Sellers," Erie (PA) *Times-News* (1 January 1978): n.p., both in Barbara Seaman Papers, Schlesinger Library, Cambridge, Ma., Acc.#82-M33–84-M82, Carton 2, Folder 91. The best-read nonfiction book in Erie County was *The Beer Can and the Beer Can Collector's Bible*; the best-read fiction book was *Coma*, by Robin Cook.

40. Barbara Seaman, *The Doctors' Case Against the Pill* (New York: Peter H. Wyden, 1969). For more on Seaman's book and the 1970 pill hearings, see Elizabeth Siegel Watkins, *On the Pill: A Social History of Oral Contraceptives, 1950–1970* (Baltimore: Johns Hopkins University Press, 1998), chapter 5.

41. Reitz, *Menopause*, 6–8.

42. Barbara Seaman Papers, Schlesinger Library, Cambridge, Ma., Acc.#82-M33–84-M82, Carton 3, Folder 111.

43. *New York Times* (8 May 1976): 13. Louis Parnish, M.D., *No Pause At All* (Binghamton, New York: Reader's Digest Press, 1976).

44. *New York Times* (15 January 1978): BR6. Louisa Rose (editor), *The Menopause Book* (New York: Hawthorn Books, 1978).

45. *New York Times* (27 February 1976): 14. Gloria Heidi, *Winning the Age Game* (New York: Doubleday, 1976).

46. Matt Clark and Mariana Gosnell, "Managing the Menopause" *Newsweek* 97 (9 February 1981): 92; "Menopause the Natural Way" *Good Housekeeping* 193 (October 1981): 275; Jane Porcino, "Menopause Discussion Groups" *Hot Flash* 6 (Fall 1987): 3; Louise Corbett, "Getting Our Bodies Back: Menopausal Self-Help Groups" *Woman Wise* 4 (1981): 2–4.

47. Paula Brown Doress and Diana Laskin Siegal, "Project Coordinators' Preface," *Ourselves, Growing Older* (New York: Simon and Schuster, 1987), xv.

48. Tish Sommers, "Foreword," *Ourselves, Growing Older,* xiv.

49. Paula Doress-Worters, personal communication (13 April 2005).

50. Linda Gannon and Jill Stevens, "Portraits of Menopause in the Mass Media" *Women & Health* 27, no. 3 (1998): 1–15.

51. Maryann Napoli, "Disease of the Week" *New Republic* 195 (1 December 1986): 17–18.

52. Tacie Dejanikus, "Major Drug Manufacturer Funds Osteoporosis Public Education Campaign" *Network News* (May/June 1985): 1, 3.

53. Fertility and Maternal Health Drugs Advisory Committee, Food and Drug Administration, Transcript, volume 2 (27 August 1984), 58–59. Records of the U.S. Food and Drug Administration.

54. Kathleen McAuliffe, "Prescription for a healthy old age" *U. S. News & World Report* 104 (23 May 1988): 72.

55. The circulation figures come from "*The Saturday Evening Post* Complete Demographic Profile as Defined by Mediamark Research, Inc., Fall 1986," 4. Thanks to Don Sutton of *The Saturday Evening Post* for providing me with this document.

56. *For the Woman Approaching Menopause* (Evansville, Indiana: Mead Johnson Laboratories, 1985), 54.

57. "Menopausal and Postmenopausal Women's Survey" *Saturday Evening Post* 258 (December 1986): following 56.

58. Cory SerVaas, M.D., "More About Estrogen Skin Patches" *Saturday Evening Post* 259 (January/February 1987): 52–55.

59. Cory SerVaas, M.D., "Is An Estrogen Skin Patch For You?" *Saturday Evening Post* 258 (December 1986): 56.

60. In August 2003, I found these surveys stowed away in boxes in the editorial offices of the *Saturday Evening Post* in Indianapolis, where they had sat untouched for 16 years. I am grateful to Wendy Braun and Georgia Ratliff at the *Saturday Evening Post* for facilitating my access to this previously unused collection.

61. The survey form asked respondents to provide their name, address, telephone number, height, weight, age, and race, and most of them complied. Every state in the union was represented, plus the District of Columbia, Puerto Rico, and Canada. A handful came from Americans living abroad, on every continent but Australia. The women who replied ranged in age from 26 to 94, with the large majority between the ages of 50 and 70. Most of the younger respondents had undergone surgical menopause, as a result of hysterectomy and oophorectomy; a significant proportion of older women had also had their reproductive organs removed. Except for a few who identified themselves as being of Asian descent, all of the respondents were Caucasian. Not a single person identified herself as black or African-American.

62. The Australian feminist Germaine Greer also published a book on menopause, *The Change: Women, Aging, and the Menopause,* in 1992.

63. Madonna's *Sex* was number ten. See Cader Books' website: http://www.caderbooks.com/best90.html.

64. Books in Print search (20 December 2001).

65. Faye Rice, "Menopause and the Working Boomer" *Fortune* 130 (14 November 1994): 203–212; Martha King, "The Baby Boom Meets Menopause" *Good Housekeeping* 214 (January 1992): 46–50.

66. Kathleen Kehoe, "Summary of ERT/HRT Coverage in Lay Press" (May 1991), Unpublished paper, National Women's Health Network, Washington, D.C.

67. "Estrogen: Deciding if it's right for you" *McCall's* 119 (February 1992): 28.

68. Patricia Lopez Baden, "Estrogen: Friend or Foe?" *Better Homes and Gardens* 74 (March 1996): 94–95.

69. A search of the Lexis-Nexis database (23 August 2001) of television and radio mentions of menopause and estrogen yielded the following number of "hits" per year: 1990–11, 1991–19, 1992–44, 1993–66, 1994–217, 1995–261, 1996–441, 1997–1000, 1998–806, 1999–439, 2000–937.

70. "CBS This Morning" (23 July 1990), CBS News Transcripts, Lexis-Nexis Academic Universe.

71. "CBS Evening News" (27 May 1992), CBS News Transcripts, Lexis-Nexis Academic Universe.

72. Melinda Beck, "Menopause" *Newsweek* 119 (25 May 1992): 71–72; Julia Kagan and Jo David, "Cosby Copes with Menopause" *McCall's* 118 (June 1991): 66.

73. The commercial was paid for by the Wyeth-Ayerst Women's Health Research Institute (Wyeth-Ayerst, now simply Wyeth, is the manufacturer of the Premarin brand of estrogen replacement therapy and the Prempro brand of combined estrogen-progestin replacement therapy). The commercial was viewed during the "ABC Evening News" on 21 January 2002. For a discussion of the implications of direct-to-consumer pharmaceutical advertising, see Adele E. Clarke, Laura Mamo, Jennifer R. Fishman, Janet K. Shim, and Jennifer Ruth Fosket, "Biomedicalization: Technoscientific Transformations of Health, Illness, and U.S. Biomedicine" *American Sociological Review* 68 (2003): 178.

74. "Power Surge," http://www.power-surge.com/educate.htm.

75. Margaret Morganroth Gullette, "What, Menopause Again?" *Ms.* 4 (July/August 1993): 34.

76. Katha Pollitt, "Hot Flash" *The Nation* 254 (15 June 1992): 808.

77. Beck, "Menopause": cover; Claudia Wallis, "The Estrogen Dilemma" *Time* 145 (26 June 1995): 47.

78. Donna B. Jeffe, Michael Freiman, and Edwin B. Fisher, Jr., "Women's Reasons for Using Postmenopausal Hormone Replacement Therapy: Preventive Medicine or Therapeutic Aid?" *Menopause: The Journal of the North American Menopause Society* 3, no. 2 (1996): 115.

79. L. A. Bastian et al., "Attitudes and Knowledge Associated with Being Undecided about Hormone Replacement Therapy: Results from a Community Sample" *Women's Health Issues* 9 (November/December 1999): 336.

80. Gullette, "What, Menopause Again?": 35.

81. For surveys of women's attitudes on menopause and hormone replacement therapy, see Wulf H. Utian and Isaac Schiff, "NAMS-Gallup Survey on Women's Knowledge, Information Sources, and Attitudes to Menopause and Hormone Replacement Therapy" *Menopause: The Journal of the North American Menopause Society* 1, no. 1 (1994): 39–48, and Patricia Kaufert et al., "Women and Menopause: Beliefs, Attitudes, and Behaviors. The North American Menopause Society 1997 Menopause Survey" *Menopause: The Journal of the North American Menopause Society* 5, no. 4 (1998): 197–202. For a discussion of the disconnection between increased media coverage and better informed consumers on the subject of women's health risks, more broadly, see Cristine Russell, "Hype, Hysteria, and Women's Health Risks: The Role of the Media" *Women's Health Issues* 3 (Winter 1993): 191–197.

82. Deborah Lupton explores the parallels and tensions between mainstream and feminist arguments about the use of hormone replacement therapy by menopausal women in her article, "Constructing the Menopausal Body: The Discourses on Hormone Replacement Therapy" *Body & Society* 2 (1996): 91–97.

83. For an interesting discussion of aging women and the youth-centered culture of postwar America, see Elizabeth Haiken, *Venus Envy: A History of Cosmetic Surgery* (Baltimore: Johns Hopkins University Press, 1997), chapter 4. See also Lois Banner, *In Full Flower: Aging Women, Power, and Sexuality* (New York: Knopf, 1992).

Part II

ORAL CONTRACEPTIVES

Women over 35 Who Smoke

A Case Study in Risk Management and Risk Communications, 1960–1989

Suzanne White Junod

Much has been written about the history of the development, approval, and the subsequent revolutionary social and medical aspects of the oral contraceptives, or "the Pill" as it is known colloquially, in recent years.[1] Extensive knowledge about the history of the Pill's approval and scientific understanding of its pharmacological properties and physiological effects, however, do not yet seem to balance out lingering suspicions about the overall safety of the oral contraceptives (OCs). This cloud of suspicion, it can be argued, is a direct result, at least in part, of the controversial and often contradictory early communications about the risks and benefits of the OCs.

The trials and tribulations surrounding the initial regulation of this class of drugs, in fact, can largely be attributed to the unique challenges posed, not so much by the oral contraceptives themselves, but by the need to understand and communicate new and emerging knowledge concerning specific risks and risk factors to a huge and growing population of women taking oral contraceptives. While precedents for such an effort were virtually nonexistent, the efforts required were unprecedented.[2] Regulators spoke about a drug whose overall risk/benefit profile was favorable and kept it on the market, but women watched the evening news which detailed seemingly contradictory information about the safety of this drug.[3] Much had to be learned about how to effectively communicate risk/benefit information to consumers as deeper knowledge of both the risks and the benefits of the oral contraceptives began to emerge from large-scale epidemiological studies—studies that had not been envisioned, much less launched, prior to FDA's approval of the first oral contraceptive, Enovid, in 1960. This chapter comprises a case study on the communication of drug risks to consumers in light of evolving knowledge about the oral contraceptives and in the context of the regulatory and administrative maturation of FDA as an institution.[4]

Important and lasting changes took place in the Food and Drug Administration as an institution during this period, transforming FDA from an underfunded and relatively obscure old-line government agency into a prominent and modern scientific regulatory agency.[5] Changes in FDA during these critical years of transition have been characterized as a shift in priorities from what have been described as enforcement or "cop" activities into a broader regulatory framework encompassing product safety.[6] Although the new framework necessitated more and better scientific expertise obtained both from within the agency as well as outside of it, it was also accompanied by a more publicly recognized role for the agency in assuring product safety.[7]

FDA often, during these years, confronted full-blown and emotionally charged consumer crises, rather than being subjected to sober scientific questioning about product safety issues. Unlike, for example, the headline grabbing, but relatively short-lived crises over cranberries, saccharin, cyclamate, and Red #2 that also took place during this period—crises that seemed to erupt and then achieve some kind of resolution and fade from media coverage—consumer concerns about the safety of the oral contraceptives became a recurrent media issue that must have, from the anxious consumers' viewpoint, never seemed fully resolved.[8] On the other hand, this very lack of resolution gave modern consumers their first exposure to the concept that drugs are not merely "safe" or "unsafe," but that they have benefits that may be offset by particular risks. Meanwhile, although FDA generally met the challenges recurrently posed by new questions concerning the safety of the oral contraceptives competently and with due diligence, given the changing times and circumstances of the agency and the era, FDA was on a learning curve regarding how to best communicate risk/benefit information via the media. FDA, in short, had to learn how to respond to risk/benefit issues, not so much in an entirely new way, but in a way that acknowledged the evolving nature of medical information, especially in what would eventually be formally christened the "post-marketing" phase of drug approval.

The first oral contraceptive, Enovid, by G.D. Searle, was approved by FDA in 1960, after extensive use as a hormonal treatment for menstrual irregularities beginning in 1958.[9] Although its approval was controversial, it met or exceeded the effectiveness specifications for other family planning products: it was less risky than the known risks posed by pregnancy, and it was much more effective than alternatives including the condom and the diaphragm.[10]

The first death reportedly associated with Enovid was that of a British woman, published in *The Lancet* in 1961.[11] This early news, however, was almost obscured by the alarming discovery that a drug available in Europe to pregnant women, thalidomide, had caused a virtual epidemic of birth defects. Although FDA had not approved the drug for marketing in the United States, Congress acted to enact the Kefauver-Harris amendments which provided, among other provisions, for the pre-market approval of future new drugs for both safety and efficacy.[12] Within months of the first report in Britain, the first fatal cases of thrombosis associated with the Pill had been reported among American women.[13] On May 21, 1962, FDA notified G.D. Searle that it needed to provide full information to the medical profession regarding a growing incidence of reports of serious and even fatal thromboembolisms associated with young women taking Enovid. Commissioner Larrick, apprised of these early serious health concerns, acted to appoint an ad hoc advisory committee of medical experts outside FDA to consider the evidence and make recommendations to FDA.[14] Advisory committees were a particular favorite of the Eisenhower administration and such ad hoc advisory committees were short-term, single issue committees, usually appointed by Congress, but in this case, convened by FDA itself in response to concerns about the safety of the oral contraceptives. In the August 4, 1962 issue of the *British Medical Journal*, the Medical Advisory Council to Britain's Family Planning Association published an advisory opening with the statement "the long-term effects of oral contraceptives cannot be entirely predicted," and reiterating that in their family planning clinics, oral contraceptives would be used as an "alternative" method of contraception in order that they could monitor long-term effects.[15] Reporters for the Washington *Star* and *Post* and the New York *Times* picked up on this warning prompting Commissioner Larrick to make public the results of FDA's own ad hoc committee recommendations on the same day, August 4, 1962. Although the committee determined that Enovid should remain available through doctors by prescription, Larrick noted that he had requested that Searle change its labeling to advise physicians only of "an apparent hazard in women over 35."[16] Meanwhile, all three reporters expressed healthy skepticism at the early, but admittedly alarming reports, but questioned what the standard of proof might be, in light of the recently discovered birth defects associated with the hypnotic drug thalidomide when taken during pregnancy.[17]

The final report of the committee was not issued until September 12, 1963. Noting that there were now 1.5 million women using oral contraceptives in the United States, the committee strongly recommended a "carefully planned and controlled prospective study" to gain more information on the reported risks.[18] In the meantime, the committee found that data did not support any increased risk for women over 35, so FDA announced that Searle would be able to note in its new labeling that this reported risk was not "statistically significant."[19]

In response to continuing and growing concerns about the safety of the Pill, FDA established its first permanent advisory committee in 1965, the Obstetrics and Gynecology Advisory Committee.[20] Charged with the review of contraceptive products and "to find out what effect, if any, oral contraceptives had on blood clotting," as well as to examine "the Pill's potential to cause cancer of the breast, cervix, and endometrium," the Advisory Committee sought to address the concerns of women's groups who charged that the agency was not taking documented evidence seriously enough. The Advisory Committee also allowed FDA to involve experienced outside physicians and researchers in the deliberative process.

The Committee's first meetings in November 1965 were held at FDA's Bureau of Medicine.[21] The committee concluded, as had the previous ad hoc advisory committee, that there was no need for immediate action to remove OCs from the market, but echoed concerns voiced by a World Health Organization scientific group that, like FDA, was not only concerned about the reported side effects, but about the lack of scientific data to assess them. There was early agreement on the need for a large, case control study on the relationship between oral contraceptives and blood clotting.[22] Although the committee's controversial conclusion that it had found no "adequate scientific data, at this time, to prove the pill *unsafe* [emphasis added] for human use," did not fully resolve the concerns of many women and physicians, the groundwork had already been laid for studies that would more definitively study OC risks and circulatory problems.[23]

Rather than remove oral contraceptives from the market, FDA drew upon more traditional regulatory strengths: its control over product labeling. FDA had exercised its steadily broadening control over product labeling in creative ways, beginning with the 1906 Pure Food and Drugs Act which originally provided that drugs with any of 11 dangerous ingredients had to list them on the label and ruled that labeling in general on food and drug products was not to be "false or misleading in any particular."[24] There was a fierce fight over where to place drug advertising controls prior to passage of the 1938 Food, Drug, and Cosmetic Act. The Federal Trade Commission (FTC) was given some "cease and desist" authority essentially bifurcating regulatory authority in the field, but the courts upheld and even extended FDA's control over labeling in the years that followed, defining it to include literature, placards, even books that accompany regulated products.[25] The new 1962 Drug Amendments transferred the regulation of prescription drug advertising from FTC to FDA and required that such advertising include information on contraindications and adverse effects.[26] The law's better known provisions required drug manufacturers to show, pre-market, the safety and efficacy of the drug for its intended use, based on scientifically conducted, adequately controlled clinical trials. Finally, the new law charged FDA with reevaluating the effectiveness of drugs marketed between 1938 and 1962 that had been solely evaluated on the basis of safety alone. FDA contracted with the National Academy of Sciences/National Research Council to review the New Drug Applications (NDAs) and FDA reevaluated them under its DESI program (Drug Efficacy Study Implementation).[27] Finally, the new law provided some of the first patient protection policies, including a provision requiring investigators to obtain "written consent" from patients given experimental drugs. Inasmuch as the oral contraceptives changed the relationship between many women and their doctors, the changes wrought by the Drug Amendments of 1962 upset, in several ways, the FDA's traditional relationship with physicians. Many physicians resented FDA's second guessing of the drugs they had available, of the presumption that they would not adequately advise their patients, and the meaning behind "written consent." Under regulation, manufacturers were required to provide a package insert with all drugs for the physician to use in advising patients.

FDA's experience using these new drug authorities was limited at the time officials learned of the rare, but sometimes fatal thromboembolisms potentially associated with use of the OCs. Traditionally, FDA had employed its control over product labeling to make physicians aware of serious drug side effects, relying on them, in turn, to adequately counsel their patients. Utilizing the agency's labeling authority in conjunction with the regulations requiring package inserts, FDA published a policy concerning the "promotional labeling" of oral contraceptives in September 1967.[28] If the contraceptive package was simply labeled in such a way that a physician would provide it to a patient along with his own oral instructions—the traditional manner of drug labeling—and had no promotional material embedded or attached, then the risks did not have to be listed on the package. However, if the package referred to safety or effectiveness, full information for the patient, in laymen's language, was to be provided, including side effects and symptoms of serious adverse effects which should be promptly reported. "For example," FDA officials stated, "it is not necessary to state that thromboembolic episodes have been associated with the use of the drug, but it is necessary to advise that such symptoms as severe or persistent headache or dizziness, leg pain or swelling, chest pain, change in vision, etc., should be promptly reported."[29] This interim position, however, was short lived.

The final reports from British researchers Inman, Vessey, and Doll were published in the British Medical Journal in 1968, and on April 29, 1968, FDA announced that manufacturers were being asked to revise their uniform labeling to reflect the findings of British researchers who had indeed found an association between the use of oral contraceptives and thromboembolic diseases.[30] The published results were in line with the preliminary results from Richard Sartwell's work at Johns Hopkins, expected within the year. Herbert L. Ley, Jr., M.D., then Director of the FDA's Bureau of Medicine, had already notified oral contraceptive manufacturers by telephone of a meeting with FDA officials on May 8 to discuss "prompt revision of uniform labeling." On June 28, 1968, physicians began to receive a "Dear Doctor" letter signed by James L. Goddard, M.D., Commissioner of Food and Drugs, acknowledging the association found between the use of the oral contraceptives and the incidence of thromboembolic disorders. New labeling, he noted, would be in use by July 1, 1968. The labeling "is uniform for all drugs in this class, and is in the standard format recommended by the FDA for all drug labeling."[31] FDA staffers had already concluded that there were 16 oral contraceptives on the market with an estimated eight million users, while 51 deaths associated with OCs had been reported to FDA or published in the literature.[32]

Early concerns that the oral contraceptives might be linked to cancer, particularly breast cancer, as well as to heart attacks, strokes, and blood clots began to make the news. Such reports, however, were often accompanied by reference to the overall benefits of these drugs.[33] Early broadcast news writers and anchors were noticeably prone to expressing appreciation for the tangible and theoretical benefits offered to society by the Pill, especially in the context of the population threat. There was little reference to any health risks that individual women taking the Pill might encounter. FDA's focus on the safety of the Pill for individual and specific women represented a somewhat specialized perspective on this issue.

In March 1969, popular Washington *Post* columnists Drew Pearson and Jack Anderson published accusations that FDA was sitting on data showing much higher than previously published fatality rates from the oral contraceptives. "The matter is so serious," they wrote, "that two Congressional committees have been quietly investigating reports that at least 10 percent of all adverse reaction reports are fatalities and that one third of the recent reports on one specific pill involve death."[34] FDA's response pointed out that the columnists were mistaken. Adverse event reports, they pointed out by their nature, were reports of *suspected,* rather than confirmed, adverse drug reactions.[35] FDA officials had examined all 9,000 adverse reports cited by Anderson and Pearson while the FDA Advisory Committee on Obstetrics and Gynecology itself had reviewed some 3,000 adverse reports *related* to oral contraceptives. The report also spelled out in numerous appendices, the steps FDA had taken to study the issue of blood clots and oral contraceptives, including Sartwell's analysis of hospital records related to idiopathic thromboembolic episodes.[36]

In the summer of 1969, FDA's Obstetrics and Gynecology Advisory Committee published its long-awaited *Second Report on the Oral Contraceptives,* in which it reported on the results of the British studies as well as Sartwell's study. The Advisory Committee estimated the risk of thromboembolism for women using the Pill to be 4.4 times that of a nonuser. The studies together, summarized the committee, suggested that the mortality from thromboembolic disorders attributable to the oral contraceptives was about three per 100,000 women per year—adding less than 3% to the total age specific mortality in OC users.[37]

Nonetheless, the report noted, "American women are sufficiently interested in oral contraception to continue its use despite some alarming reports in the national press." They reported twenty different brands on the market, including combined and sequential formulations, which was twice the estimate of users found in a 1965 fertility survey. "This apparent doubling of the numbers," noted the Advisory Committee, "suggests a much wider use among older women and those of limited education. Such a trend could be forecast from the increased availability of contraceptive services in many of the poorer areas of our big cities."[38]

As Elizabeth Watkins, Andrea Tone, Lara Marks, and others have discussed, the hearings convened by Senator Gaylord Nelson to hear testimony concerning the safety of the oral contraceptives, beginning on January 14, 1970, marked the true turning point in early debates and discussions over the safety of the Pill.[39] It also marked a turning point in media coverage of this issue. Long gone were the reassuringly optimistic staged assessments from government officials. Network news coverage of the Nelson Committee hearings in January 1970 was revolutionary and riveting. The first female reporters, including Barbara Walters, appeared on camera covering Senator Gaylord Nelson's hearings on the safety of the Pill. The hearings made exciting television as women from all walks of life, including young women in mini-skirts, mothers with babies, and middle-aged female scientists and physicians converged on the Senate with demands to be heard. Initially rebuffed, they were eventually allowed to testify in a second series of hearings. As Elizabeth Watkins has noted, after the first round of televised hearings, "the television-viewing public knew a great deal more about the controversy surrounding the safety of the pill, but no more about whether the pill was safe to take."[40]

During a second round of hearings, FDA Commissioner Dr. Charles Edwards testified concerning the safety of the Pill, and extended an olive branch to the outraged women's groups at the hearings when he endorsed and announced that FDA would require that manufacturers include printed warnings about side effects from the Pill in birth control packages. However, problems soon arose as controversy ensued over the contents and wording of the warning. This controversy soon engulfed not only FDA, physicians' groups including the AMA, and pharmacists, but the Department of Health, Education, and Welfare, the pharmaceutical industry, and women's advocacy groups as well.[41]

Although the first proposed patient package insert was a detailed four-page leaflet, in the end, the first patient package insert for the oral contraceptive was considerably shorter and less comprehensive.[42] It was provided, not to patients, but to physicians to share with their patients upon request. Nonetheless, the fact that the patient package insert (PPI) was adopted at all was quite significant.[43] In rejecting the arguments of

Cigarette smoking increases the risk of serious adverse effects on the heart and blood vessels from oral contraceptive use. This risk increases with age and with heavy smoking (15 or more cigarettes per day) and is quite marked in women over 35 years of age. Women who use oral contraceptives should not smoke.

Figure 1.

physicians, pharmacists, and drug manufacturers claiming that the PPI represented an illegal intrusion into the practice of medicine and the physician-patient relationship, the government adopted the position that it was acting to protect patients precisely because physicians were not providing their patients with "the needed information in an organized, comprehensive, understandable, and handy-for-future-reference form." The oral tradition of instructing patients, the government maintained, could not adequately protect patients. "The necessary information for the safe and effective use of oral contraceptives is too complex to expect the patient to remember everything told her by the physician."[44] All three networks covered the disputes over the warnings to be dispensed with the Pill.[45]

The links between oral contraceptive use and thromboembolic events were fairly well known by the end of the hearings. Fortunately, the incidence of serious thromboembolic events remained low and the actual mortality was estimated to be 3 in 100,000 users. As lower estrogen dose OCs became the norm rather than the exception, these mortality rates dropped even lower. A survey of ob-gyns conducted after the hearings in 1970 reported that in spite of the adverse publicity, the overwhelming majority of physicians continued to prescribe oral contraceptives.[46]

Ironically, shortly after the Nelson hearings ended, British researchers published some good news. In April 1970, the British Committee on the Safety of Drugs found that the incidence of deep vein thrombosis in the leg was reduced about 25% if 50 microgram (mcg) estrogen formulations were prescribed instead of higher doses.[47] Ongoing research had indicated that the risks of blood clots, heart attack, and stroke were directly related to the amount of estrogen in the various versions of the Pill. Research had sustained earlier findings showing that contraceptive effectiveness could be maintained with only 50 mcg of estrogen and 7.5 million women were using oral contraceptives.

Complicating almost all discussions of the Pill's safety during this period was the fact that the oral contraceptives were available in several different strengths by every manufacturer. The earliest pill formulation in the United States, Enovid, approved in 1960, had 150 micrograms of estrogen (mestranol). Because of its high incidence of side effects, Enovid 10 mg soon gave way to Enovid 5 mg and an entire series of so-called monophasic pills, all approved between 1960 and 1968. The most popular brands were Enovid and Ovulen (G.D. Searle), Ortho Novum (Ortho and Johnson & Johnson), Norinyl (Syntex), and Norlestrin (Parke Davis/Warner Lambert). The most widely used formulations in each brand contained between 80 and 100 micrograms of the estrogen component (as either ethinyl estradiol or mestranol).

In the mid-1960s, a series of "sequential" oral contraceptives received FDA approval and these formulations also contained 80 to 100 micrograms of the estrogen component. The Royal College of General Practitioners' Oral Contraception Study, an ongoing cohort study begun in 1968 to follow Pill users, reported, early on, that the sequentials had too little progestin, which offered no advantages but did seem to create some increase in uterine cancer.[48] The sequentials, therefore, were removed from the U.S. market in the mid-1970s, but by the late 1960s, there had already been a major shift in prescribing patterns. In this second phase, dating roughly from the late 1960s to the mid-1970s, almost all of the oral contraceptives by every manufacturer contained 50 micrograms of the estrogen component.

A third phase in drug development and prescription patterns was characterized by the emergence of lower-dose 35 microgram oral contraceptive formulations. Most of the manufacturers of these brands received approval between 1973 and 1989 and they dominated the oral contraceptive marketplace throughout the 1970s. The fourth and latest period in prescribing patterns emerged in the 1980s and 1990s, characterized by an increasing number of options including, most recently, a new delivery system for the oral contraceptive via a transdermal patch. Most of the new products contained new estrogen and progesterone combinations, but the estrogen component ranges more widely—from 20 to 30 micrograms of estrogen (see Appendix).

Five days before the official publication date of the findings of the British Committee, a consumer, Carolyn Morgan, had requested in writing detailed information on birth control pills, including brand names and estrogen doses from the Department of Health, Education, and Welfare. Soon afterward, her request was denied. The Freedom of Information Act, first enacted in 1946, required each agency to establish its "procedures, opinions, policies, and rules" and to make records not exempted available to requestors as soon as possible. FDA's rules specifically exempted its "trade secret and confidential information." It also included internal documents and inter- and intraagency memoranda.[49] The Morgan suit was filed with the assistance of Ralph Nader's Health Research Group, with the broader aim of opening up more of FDA's records. FDA's exclusion of almost all its records under the "trade secret" provision of the law had made it a target for the wrath of women such as Morgan, who wanted to know how to safely take the oral contraceptives. In hearings that spanned six weeks between March 6 and April 19, 1972, the House Committee on Government Operations held hearings on the administration of the Freedom of Information Act. FDA was heavily criticized during the course of the hearings and by year's end, FDA had a new set of extensive regulations opening up the vast majority of FDA's previously exempted files. Authored by FDA Chief Counsel, Peter Barton Hutt, the new regulations predated, by almost a year, the passage of extensive amendments to the Freedom of Information Act in 1974, which reaffirmed, updated, and codified the commitment to disclosure and openness in the release of federal records. Although the Morgan case was appealed, Ms. Morgan was ultimately unsuccessful in getting the New Drug Applications that she sought, but this case and others did open up more agency documents for scrutiny by consumers as well as industry.[50]

Change in the agency's information policy was just one of a series of new initiatives begun under Edwards, and continued and refined under subsequent leaders, which sought to accommodate the needs of consumers for more openness in its dealings with the Food and Drug Administration. In 1970, a new Office of Consumer Affairs was created in the agency and in 1972, FDA's publication *FDA Papers* was renamed *FDA Consumer* and began to address directly issues of interest to consumers as opposed to industry, including "one of the most talked about medications in the past 50 years—"the Pill."[51] Advisory Committees brought more openness into the agency's deliberations, especially as sunshine laws began to require public meetings and opportunities for public comment. The committees themselves were increasingly sophisticated, drawing members widely from both industry and from nonprofit consumer organizations. Commissioner Edwards himself chaired the first meeting of FDA's National Drug Advisory Committee (NDAC) on April 20, 1972, and the new findings on lower dose oral contraceptives were included on the agenda. Conceived of as a kind of "science court," in which participants could sharpen their understanding of issues and become educated in a neutral forum under the tutelage of experts, they offered a new opportunity for open discussion on important issues of the day and illustrated the agency's willingness to explore new forms of communication.[52]

Charles Edwards left FDA to accept an assistant secretary position in HEW, where he could still interact with FDA through the newly appointed commissioner, Dr. Mack Schmidt, a physician and an academic administrator from the University of Illinois. During Schmidt's tenure, safety issues surrounding the oral contraceptives seemed relatively settled and most outstanding issues were settled quietly through revisions in the package labeling and the patient package inserts.[53] In 1973, for example, FDA ordered physicians and suppliers to supply information in the PPI concerning the side effects of oral contraceptives when prescribed as a "morning after" birth control method.[54] This may well have been the first time many women were made aware of the fact that the Pill could be taken "after the fact" to prevent pregnancy. In 1974, noting a rise in venereal diseases attributed to oral contraceptive use in lieu of barrier methods of contraception, FDA and CDC agreed to request a new statement in the drug's packaging and PPI: "Caution: Oral Contraceptives are of no value in the prevention or treatment of venereal disease."[55]

During the relative media calm following the Nelson hearings, women had to assess and reassess not only their own contraceptive needs but also their relations with the physicians who prescribed them. It was up to individual physicians to discuss with their patients and implement the recommendations for prescribing lower estrogen dose contraceptives. Meanwhile, scientific evidence was beginning to accumulate on a broader cardiovascular threat to health posed by the oral contraceptives. Physicians had learned from their own patients' experiences that steroids, most notably prednisone and cortisone, could create serious and sometimes irreversible side effects with prolonged or daily use, leading them to be especially vigilant for minor side effects that might be early indicators of more serious problems. Early reports of migraines, altered vision, changes in libido, blood pressure, and glucose tolerance all seemed potentially alarming, but were soon shown to be either minor, not related to Pill usage, or of short duration as the patient adapted to the drug.[56] Likewise, some of the early published studies on the OCs had noted a high incidence of myocardial infarction (MI) among users, but had concluded that the relationship was uncertain.

Vessey and Doll's landmark study on thromboembolism, for example, noted that thromboembolic patients were, on average, heavier cigarette smokers than the control patients, but they did not find a statistically significant difference.[57] As had been the case with thromboembolic events, no one was sure what the baseline rate for such serious events in healthy young women was naturally since they were considered so rare that researchers had had no previous cause to study the issue. Kay reported from the Royal College of General Practitioners' Study in *Lancet* that preliminary analyses of data based on 32,000 women had shown that oral contraceptive users were more likely to be heavy smokers than nonusers. Their conclusion was that the possible influence of smoking habits had to be assessed before concluding that adverse events were attributable solely to the effect of oral contraceptives.[58]

A year later, researchers had postulated at least four ways oral contraceptives might aggravate coronary occlusions: raising cholesterol levels by reducing carbohydrate tolerance, leading to raised blood pressure, and increasing platelet adhesion.[59] Correspondents to the *British Medical Journal* soon pointed out results from a U.S. Public Health Service study that they felt supported a possible potentiating role for smoking as a cause of thromboembolic events. Retabulating the data from the Vessey and Doll study, they found that the heaviest smokers also had the most thromboembolic events.[60] Finally, in the same April 25, 1970 article that had shown beneficial reductions in risk from lower estrogen dose pills, the British Committee on the Safety of Drugs had also affirmed that oral contraceptives clearly played "a definite but small part" in the total range of factors contributing to cerebrovascular disease. "We must still determine," they concluded, "how small that part is and the precise mechanism through which it occurs."[61]

Meanwhile 1975 marked the fifteenth anniversary of the approval of Enovid for contraception. An article in *JAMA* published in February that year warned of a statistical link between the Pill and an increased risk of stroke and recommended that women with migraines, high blood pressure, or heavy smokers be made aware of the dangers. The article's conclusions were noted by ABC and CBS, but did not receive widespread coverage in the broadcast news.[62] In fact, much of the published news as well as the broadcast news on the Pill was reassuring. CBS ran its longest report on the oral contraceptives, a full four minutes, on May 9, 1975.[63] Articles pointed out to women that the lower estrogen dose OCs had fewer annoying side effects than the older high dose pills, and that in addition to their contraceptive effects, they offered other advantages "including relief of cramps or painful menstruation, a smaller volume and shorter duration of monthly bleeding, greater cycle regularity, less premenstrual tension, and a "feeling of general well being." Reduced menstrual bleeding, the authors noted, could be especially useful for malnourished or anemic women.[64]

The storm broke on May 3, 1975, when the results of two long-awaited case control studies under the auspices of the Royal College of General Practitioners were published in the *British Medical Journal.*[65] For the first time since the oral contraceptives had been marketed, all three networks covered the same story on the oral contraceptives on the same night— August 27, 1975. The time which had elapsed between publication of the data and the televised broadcast warning had been spent by the FDA Ob-Gyn Advisory Committee members, agency experts, and public health officials, assessing the data, deciding on a course of action, and drafting new label regulations for oral contraceptive packages. The British reports had concluded that women over 40 who continued to take oral contraceptives were five times more likely to suffer a heart attack. FDA announced, sooner than it had originally planned, that it was drafting new warning labels for birth control pills.[66] FDA's on-camera spokesperson, Dr. Marion Finkel, Director of the Office of New Drug Evaluation, spoke on NBC carefully explaining that the studies had shown that although there was some slight risk of heart attack in women aged 30–33 on the Pill, it was so small that studies had not been possible because actual heart attacks were so rare in that age group. The real risk, the studies seemed to indicate, said Finkel, came after age 40. Therefore, FDA was recommending that women over 40 consider getting off the Pill.[67]

FDA explained the problem to its own employees by noting that "the increased risk of thromboembolism, pulmonary embolism, and cerebral thrombosis in oral contraceptive users had already been established," but that the data on the risk of coronary thrombosis had been equivocal.[68] The "most important conclusion" from the first new British study, however, was the fact that oral contraceptive use was an independent risk factor for coronary thrombosis.[69] In other words, it was not explained by concomitant risk factors. The greater the number of risk factors for coronary artery disease (oral contraceptive use, smoking, diabetes, history of preeclampsia toxemia, hypertension, obesity, high blood lipids) the higher the risk of developing an MI. The effects, moreover, were synergistic (multiplied) rather than additive. Based on the two studies, FDA assessed the overall annual additional risk of developing a fatal or nonfatal MI to be about 1 per 1,000 in women over 40.[70] It would be even higher in women with additional risk factors, and at that point in time, nearly 30% of women were considered to be smokers.[71] Therefore FDA was acting upon the recommendations of its Obstetrics and Gynecology Advisory Committee, and had revised oral contraceptive labeling to reflect that women over 40 faced increased risks from oral contraceptives, and recommending that women over 40 use other forms of contraception.

By mid-October, when FDA's new warning labels were released, CBS's Walter Cronkite spoke for 30 seconds about news that taking the Pill after age 40 increased the risk of heart attack, stroke, and blood clots.[72] ABC, however, devoted a full two minutes to an extended discussion of the issue. "The Pill is safe for the vast majority of women using them, but you should not use the Pill if you're over 40 due to the increased risk of blood clots and stroke." Medical reporter David Culhane ended by anticipating the reaction they must have feared most, "Before you panic, talk to your doctor about risks in light of what he knows about you and your lifestyle."[73]

As indicated by these network news broadcasts, 1975 marked an important turning point in press coverage of the Pill. News reports had certainly reported safety concerns about the Pill as well as other news related to its societal effects, changes in lifestyle brought about by the Pill, and controversies over labeling and information provided to consumers about the oral contraceptives over the years, but with news about these broadened risks to women over 40 taking oral contraceptives, the evening broadcast news reports shifted their focus almost exclusively to concerns about the safety of the Pill.[74]

For the first time in its 15-year vigilance over the oral contraceptives, FDA itself appeared to waver in its overall assessment of the safety of OCs, as the first fruits of large-scale research trials yielded solid, yet incontrovertible evidence that most oral contraceptive users had far less risk of experiencing a major, but rare, thromboembolic event, than they did of experiencing a more common, but equally serious MI or heart attack. Dr. Richard Crout, FDA's Director of the Center for Drug Evaluation and Research, summed up his opinion on the Pill in a 1976 CBS interview by saying that women "must approach it with great care" and prophesized that the oral contraceptives would "disappear when an equally effective method of birth control was discovered."[75] This seemed to be a telling concession from a regulator whose agency continued to maintain that the Pill was safe enough to remain on the market.

Shortly thereafter, while Britain took a pragmatic approach, warning women over 40, especially those who smoked, that that they faced additional dangers, FDA acted to strengthen labeling requirements for oral contraceptives in general.[76] FDA announced that women would receive a detailed brochure, describing the risks and benefits of the Pill with every prescription and every refill, and the *Federal Register* notice itself contained nine pages detailing the new warnings and proposed warnings. Instead of being distributed through physicians, the brochures would be supplied by the person dispensing the drug, usually the pharmacist. A press release accompanying the announcement contained was tersely worded.

> FDA neither advocates nor discourages the use of birth control pills, but believes that women should have this choice of contraceptive method available to those who want to use it. The selection of a contraceptive is not like the selection of other drugs. Contraceptives are not intended to alleviate or cure diseases. The choice of a contraceptive is a personal one and the selection must be based on accurate, balanced information presented in a form that can be understood.[77]

In 1976, CBS aired a three-part "Special Report" on the Pill. The first segment alone was six minutes and the series represented the longest and most comprehensive report on the Pill to air on network news to date. Walter Cronkite noted that the Pill had been used by over 30 million women since its introduction and remained the most effective means of birth control on the market. "However," emphasized Cronkite, "slowly over the years, adverse events have come to the attention of women in bits and pieces. The result?: confusion about the risks and benefits of an important medical product."[78] Part 2 in the CBS "Special Report," aired the following night, was highlighted by a clip of Miss America, 1963, Mrs. Jacqueline Townsend, who had suffered a stroke in 1970 after "resuming" birth control pills. Although her ability to speak had returned after three to four years, she urged women not to take the Pill.[79] Finally, Part 3 concluded the series and largely cited statistical evidence concerning the Pill's safety. "Overall estimates are that 1 in 100,000 women taking the Pill suffer death and disability as a result," reported David Culhane. His explanation of the risks to women over 40 was both simple and understandable. Displaying a chart with clearly marked curves, he showed that up to age 40, pregnancy and childbirth were more risky than the Pill, but after age 40, the curve shifted, making the Pill more risky.[80] New statistics showed that although 36% of women were taking the Pill, barrier methods were rising in popularity after many years of decline (19%), and sterilization was, in fact, the fastest growing method of birth control (24%).[81]

While Britain had expressed particular concern about the results of the Royal College of Practitioners' findings that dealt with women who smoked, U.S. regulators, at least at first, were more concerned about the same studies' findings of the dangers to women over 40, in general. Soon after the report, however, FDA's perspective changed. The change was more of a change in emphasis as it coincided with an intense anti-smoking campaign initiated in January 1978 by Health and Human Services Secretary Joseph Califano. The initiative included a ban on smoking on commercial airplanes, a new Surgeon General's Report, more restrictions on smoking in government buildings, insurance premium discounts for nonsmokers, and an increase in the excise tax on cigarettes. An essential part of the new war was going to be "new warnings" on birth control pills warning women that if they smoked, they increased their risk of heart attacks dramatically.[82] The problem with the strategy, however, was soon evident. Given a choice between smoking and taking the Pill, most women found it easier to quit taking the Pill than to quit smoking.[83]

When FDA's final rule on new labeling for the oral contraceptives were published, on January 31, 1978, the new rules significantly expanded the amount of information required to be provided to patients, as well as the revisions on how the PPI was to be prepared and distributed. The most notable addition to the regulation, however, was a provision for a "black box" warning, stating that "cigarette smoking increases the risk of serious cardiovascular side effects in oral contraceptive users." The discussion in the section entitled "Dangers of Oral Contraceptives" was revised to reflect the risks of serious cardiovascular side effects including heart attacks.[84]

ABC does not seem to have covered Califano's crusade until January 24, 1978, when Barbara Walters announced that after April 3, oral contraceptives would carry a new warning and pharmacists would provide women picking up a prescription for the Pill with a brochure acknowledging the increased risk of serious adverse side effects if they smoked while on the oral contraceptive.[85] CBS, however, and its medical reporter, Richard Roth, were the first to show the new specific Pill package warning. [86]

FDA Commissioner Donald Kennedy, interviewed on the new labeling initiative, remained concerned by the British findings. Noting that estimates were that 30–40% of women on the Pill also smoked, he emphasized that they had "ten times the risk of cardio-vascular incidents if they smoked and took oral contraceptives," which puts over "4 million women at risk" in the United States alone. Although the risk was substantial, Kennedy reiterated that this risk was still smaller than the risk of complications from childbirth. This might normally have ended a news interview, but in a clear sign that the media was echoing and expressing women's frustration with the piecemeal pattern of safety warnings on the Pill, Kennedy was asked, "What—as a private citizen—would you recommend to your own daughter or someone you cared about?" Kennedy's response was cautious. "As a private citizen, I would tell them to find another method of contraception," he stated simply.[87]

The interview with Kennedy, following the unveiling of the black box warning, was an important turning point in the late 1970s. Both the charts and the black box warning offered women and their physicians new guidance on the risks of cardiovascular disease associated with Pill use. Until smoking was separated out from the cardiovascular data, it had appeared that the cardiovascular risk was largely confined to women over 40, when, in truth, the incidence of cardiovascular disease in women began a steady climb upward throughout their reproductive years. The point on the graph at which the danger of taking the Pill exceeded the well-established dangers of childbirth was around age 40 for most women. Women who smoked, however, skewed the chart. By plotting the point at which the dangers of the Pill exceeded the protection it offered for smokers separately, the risks clearly outweighed the benefits—but at 35, not 40. As the estrogen content in low dose pills continued to decline and more women switched to the lower doses, and older women who smoked stopped taking the Pill, the safety profile for women taking oral contraceptives improved notably.

At last there was a coherent explanation and a shorthand reminder of the observational and experimental data—at least as far as heart disease was concerned.[88] The oral contraceptive should not be used by women over 35 who smoked. The warning was simple, it neatly encapsulated the scientific data, and it did not require women to listen to complicated

DOCTOR: Just don't do it.
Smoking increases the risks,
especially if you're over 35.

Figure 2.

explanations of risks and benefits from their physicians or seek them out on their own from brochures, *Federal Register* notices, or the ever-lengthening patient package inserts for oral contraceptives. It was an important line of demarcation, establishing a clear and easily remembered delineation between safe and unsafe users of oral contraceptives. It remains one of the oldest and most recognizable of women's health warnings into the twenty-first century, as illustrated by this 2005 storyboard from a television advertisement for the Ortho-Evra contraceptive patch, in which the physician talks directly to the patient about the risks posed by the oral contraceptives, noting that "Smoking increases the risk [of heart attack, stroke, and blood clots] especially if you're over 35."[89]

Once the warning had become a commonplace benchmark for discussion, slight modifications to the scientific findings did not significantly threaten the new status quo. In 1982, a new version of the Pill, known as the "biphasic" pill, was introduced (see Appendix). In 1984, Kay and his Royal College colleagues updated the original findings, finding a slightly reduced rate of relative risk.[90] By 1985, a new triphasic pill was approved (see Appendix). In contrast, by 1986, high-dose oral contraceptives made up only 3.4% of the oral contraceptive market. Nonetheless, some 400,000 women were still using high-dose estrogen pills, but in 1988, the three drug companies still making high-dose estrogen oral contraceptives voluntarily withdrew all remaining products over 50 mcg estrogen.

Practicing physicians increasingly began to think of the Pill in a new way—as an excellent means of birth control for young women.[91] With the new understanding of the Pill's cardiovascular effects, it became clearer that the Pill was still the best method of contraception in a young woman's reproductive life when her cardiovascular risks are lowest.

In its eternal quest to keep labeling information up-to-date, FDA proposed new revisions for patient package inserts for oral contraceptive drug products in 1987. The new proposed changes, which required only two pages in the *Federal Register,* were intended to simplify both the format and the content and make the insert more understandable as well as make it easier to keep updated. Instead of specific wording requirements, information in the PPI would appear as a list of general categories of information.[92] The changes were finalized in 1989.[93]

In 1989, FDA approved the use of low-dose oral contraceptives for healthy, nonsmoking women over 40.[94] Kay and his colleagues even recast their own conclusions in a more positive light, concluding that "nonsmokers using low-progestogen dose brands may safely use oral contraceptives, probably up to the age of 45 years."[95] Although even low-dose oral contraceptives still pose some cardiovascular risks, those risks are much better understood than they once were.[96]

The oral contraceptives and discussions about the evolving knowledge concerning their risks and benefits have been at the forefront of many changes and innovations in FDA's communications with consumers. Just as women were instrumental in the passage of the original 1906 Pure Food and Drugs Act as well as passage of the 1938 Food, Drug, and Cosmetic Act, women continued to influence FDA in important ways even after FDA's enabling legislation was in place.[97] Concerns about the safety of oral contraceptives necessitated creation of the first permanent agency Advisory Committee and inspired activist feminist leaders to push their way into the earliest closed proceedings. The first patient package insert was demanded by women and although it did not meet all of their require-ments, it broke new regulatory ground as FDA reached beyond physicians and attempted to ensure that what they gave patients was accurate and communicated real and up-to-date information about the risks they assumed in taking oral contraceptives. Requests for information under the Freedom of Information Act helped open up some agency records and motivated women to learn more about the medicines they and their fami-lies took. FDA, in turn, learned the importance of providing accurate and timely communication of risks to consumers and the importance of tele-vision as a critical component of modern communication.

APPENDIX

Abbreviations

Progesterone Component	Estrogen Component
NET - Norethindrone	EE - Ethinyl Estradiol
NEA - Norethindrone Acetate	ME - Mestranol
NRG - Norgestrel	
LNG - Levonorgestrel	
DSG - Desogestrel	
NGS - Norgestimate	
EDA - Ethynodiol Diacetate	
DMT - Dimethisterone	
CMA - Chlormadinone Acetate	
MPA - Medroxyprogesterone Acetate	

Note: Variants of the names below may include terms for days of use (21 or 28) and/or the addition of iron (FE or FerrousFumarate)

1A.   MONOPHASIC
> 50 MCG ESTROGEN

Manufacturer	Year	Brand Name	Progestogen (Mg)	Estrogen (Mcg)
Searle	1960	Envoid 10 Mg	9.85 NET	150 ME
	1961	Envoid 5 Mg	5.0 NET	75 ME
	1964	Envoid E	2.5 NET	100 ME
	1966	Ovulen	1.0 EDA	100 ME
Ortho	1963	Ortho Novum 10 Mg	10.0 NET	60 ME
(Rw Johnson Pri)	1963	Ortho Novum 2 Mg	2.0 NET	100 ME
	1968	Ortho Novum 1/80	1.0 NET	80 ME
Syntex	1964	Norinyl 2 Mg	2.0 NET	100 ME
	1968	Norinyl 10 Mg	10.0 NET	60 ME
	1968	Norinyl 1/80	1.0 NET	80 ME
Parke Davis (Warner Lambert)	1967	Norlestrin 1mg	0.5 NEA	100 EE

1B. SEQUENTIAL
>50 MCG ESTROGEN

Manufacturer	Year	Brand Name	Progestogen (Mg)	Estrogen (Mcg)
Ortho	1965	Ortho Novum Sq	14x 0 NET	80 ME
			6x 2.0 NET	80 ME
Syntex	1967	Norquen	14x 0	80 ME
			6x 2.0 NET	80 ME
Lilly	1965	C Quens	15x 0	80 ME
			5x 2.0 CMA	80 ME
	1969	Estalor	14x 0	100 ME
			7x ? CMA	100 ME
Mead Johnson	1965	Oracon	16x 0	100 EE
(Bristol Myers Squibb)			5x 25.0 DMT	100 EE

2. MONOPHASIC
50 MCG ESTROGEN

Manufacturer	Year	Brand Name	Progestogen (Mg)	Estrogen (Mcg)
Searle	1970	Demulen	1.0 EDA	50 EE
Syntex	1969	Norinyl 1/50		
		Noriday	1.0 NET	50 ME
Ortho	1967	Ortho Novum 1 Mg	1.0 NET	50 ME
		Ortho Novum 1/50		
		Ovanul 1/50		
Parke Davis	1964	Norlestrin 2.5 Mg	2.5 NEA	50 EE
	1968	Norlestrin 1/50	1.0 NEA	50 EE
Mead Johnson	1976	Ovcon 50	1.0 NET	50 EE
Wyeth	1968	Ovral	0.5 NRG	50 EE
(Wyeth-Ayerst)				
Lederle	1973	Zorane 1/50	1.0 NET	50 EE
Upjohn	1964	Provest	10.0 CMA	50 EE

3. MONOPHASIC
<50 MCG ESTROGEN

Manufacturer	Year	Brand Name	Progestogen (Mg)	Estrogen (Mcg)
Searle	1981	Demulen 1/35	1.0 EDA	35 EE
Ortho	1974	Ortho Novum 1/35 Neocon	1.0 NET	35 EE
	1974	Modicon	0.5 NET	35 EE
	1989	Ortho Cyclen	0.25 NGS	35 EE
	1992	Ortho Cept	0.15 DSG	30 EE
Syntex	1974	Brevicon 1 Mg	1.0 NET	35 EE
	1974	Brevicon	0.5 NET	35 EE
	1986	Norquest	2.0 NET	35 EE
Parke Davis	1973	Loestrin 1/20	1.0 NEA	20 EE
	1976	Loestrin 1.5/30	1.5 NEA	30 EE
Mead Johnson	1976	Ovcon 35	0.4 NET	35 EE

3.   MONOPHASIC
<50 MCG ESTROGEN   (CONT.)

Manufacturer	Year	Brand Name	Progestogen (Mg)	Estrogen (Mcg)
Wyeth	1975	Lo Ovral	0.3 NRG	30 EE
	1982	Nordette	0.15 LNG	30 EE
Lederle	1973	Zorane 1.5/30	1.5 NET	30 EE
	1989	Zorane 1/20	1.0 NET	20 EE
Organon	1992	Desogen	0.15 DSG	30 EE

4.   MULTIPHASIC
<50 MCG ESTROGEN

Manufacturer	Year	Brand Name	Progestogen (Mg)	Estrogen (Mcg)
Ortho	1982	Ortho Novum 10/11	10x 0.5 NET 11x 1.0	35 EE
	1984	Ortho Novum 7/7/7	7x 0.5 NET 7x 0.75 7x 1.0	35 EE
	1984	Ortho Novum 7/14	7x 0.5 NET 14x 1.0	35 EE
	1992	Jenest Ortho Tricyclen	7x 0.180 NGS 7x 0.215 7x 0.250	35 EE
Syntex	1984	Tri Norinyl	7x 0.5 NET 9x 1.0 5x 0.5	35 EE
Wyeth	1984	Tri Phasil	6x 0.05 LNG 5x 0.075 LNG 10x 0.125 LNG	30 EE 40 EE 30 EE

5.   PROGESTOGEN ONLY

Manufacturer	Year	Brand Name	Progestogen (Mg)	Estrogen (Mcg)
Ortho	1973	Micronor	0.35 NET	
Syntex	1973	Nor Qd	0.35 NET	
Wyeth	1973	Ovrette	0.075 NRG	

NOTES

Funds for the purchase of taped transcripts of evening news broadcasts from the Vanderbilt Media Archives were provided to the author through a grant from the FDA's Office of Women's Health. The support of Dr. Susan Wood, former head of that office is gratefully acknowledged. Research support notwithstanding, any opinions expressed in this chapter, as well as any errors and omissions are solely those of the author and not the FDA. Drug information appearing in the Appendix was carefully compiled by Dr. Bruce Stadel, retired medical officer in the Division of Metabolic and Endocrine Drugs, FDA Center of Drug Evaluation and Research. I thank him for his generosity in allowing me to publish his compilations. I am also indebted to colleagues Dale Smith, Susan Speaker, Phil Teigen, and David Cantor for their comments on early drafts of this chapter. Media citations in this chapter are keyed to a typescript index in the FDA History Office. Able research assistance was provided by Cindy Lachin in the FDA History Office.

1. Elizabeth Watkins, *On the Pill: A Social History of Oral Contraceptives, 1950–1970* (Baltimore, MD: Johns Hopkins University Press, 1998); Andrea Tone, *Devices and Desires: A History of Contraception in America* (New York: Hill and Wang, 2001); Lara Marks, *Sexual Chemistry: A History of the Contraceptive* Pill (New Haven: Yale University Press, 2001); Donald Critchlow, *Intended Consequences: Birth Control, Abortion, and the Federal Government in Modern America* (New York: Oxford University Press, 1999); Patrick Vaughan, *The Pill on Trial* (London: Weidenfeld and Nicolson, 1971); Bernard Asbell, *The Pill: A Biography of the Drug that Changed the World* (New York: Random House, 1995); Marcia A. Meldrum, "'Simple Methods' and 'Determined Contraceptors': The Statistical Evaluation of Fertility Control, 1957–1968," *Bulletin of the History of Medicine* 70 (1996): 266–95; Suzanne Junod, "Perspectives on the Pill: An Essay Review," *Bulletin of the History of Medicine* 57 (2002): 333–339.

2. Suzanne White Junod and Lara Marks, "Women's Trials: The Approval of the First Oral Contraceptive Pill in the United States and Great Britain," *Journal of the History of Medicine and Allied Sciences* 57 (2002): 117–160.

3. In contrast to the changing media portrayals of the Pill's safety, especially as they applied to subgroups of women, FDA's own overall assessment of the oral contraceptives' safety as a drug class stabilized in the years following the initial controversies over its safety. FDA's Center for Drug Evaluation and Research (CDER) submitted this official summary statement for FDA's *Quarterly Activities Report* in the third quarter of 1995: "Over the years, the pill has remained a very effective and relatively safe method of contraception. Serious but infrequent adverse events that have been associated with the pill include heart attack, stroke, and thromboembolism. Most of the adverse events have been related to the estrogen in a dose-dependent manner. However, pills on the market today contain a

significantly lower concentration of estrogen than earlier pills, making these side effects less common. Furthermore, there is strong evidence that the pill has a protective effect against ovarian and endometrial cancer." This statement reflects what FDA review scientists consider to be key elements in a "good" drug approval with a favorable risk/benefit profile that benefits patients: Knowledge of the key components of the drug, their mechanism of action, the drug's important side effects, both positive and negative, and an understanding of how the side effects relate to the pharmacology of the drug.

4. Recently historians have begun paying more attention to media and communications issues in medicine. See Janet Golden, *Message in a Bottle: The Making of Fetal Alcohol Syndrome* (Cambridge: Harvard University Press, 2005); Virginia Berridge and Kelly Loughlin, eds., *Medicine, the Market, and the Mass Media: Producing Health in the Twentieth Century* (London : Routledge, 2005).

5. This is the consensus of a number of writers and observers, although different writers emphasize different elements of the transformation. See, for example, Peter Barton Hutt, videotaped presentation on "Historical Origins of Administrative Law," FDA History Office; Barbara R. Troetel, "Three Part Disharmony: The Transformation of FDA in the 1970s" (Ph.D. dissertation, City University of New York, 1996); Suzanne White Junod, "Grit to Relieve Gridlock: The 1973 Rulings on Drug Effectiveness," FDLI *Update* 5(November 1999): 9; FDA Oral History Interview, Bob Tucker and Suzanne White Junod with Richard Merrill, July 27, 2004; November 12, 2004, Washington D.C., National Library of Medicine, Bethesda, MD. Hereafter cited as Merrill Interview.

6. FDA's regulatory philosophy put the burden on industry to fully comply with the law, supported by "advisory" guidance from FDA officials. Industry often, therefore, spent a great deal of its own resources divining FDA's actions and intentions but rarely challenging them. Compliance could be daunting to industry and was entirely opaque to consumers. It sustained, however, a strong, competent, and growing legal specialty in food and drug law, while being highly cost effective for taxpayers. For an example of FDA's traditional enforcement and compliance philosophy, see Sam D. Fine and Walter R. Moses, "Advisory Opinions: A Historical Perspective," *FDA Papers* 6:1 (February 1972): 9.

7. Most recently, FDA has taken to describing its role as that of "protecting and promoting" public health. Commissioner Schmidt credits Commissioner Charlie Edwards (12/13/69–3/15/73) with raising and answering the important question about what business FDA was in during his tenure. And the answer was not, "we're in the business of putting people in jail ... it's a public safety agency. It's insuring safe and effective products on the market." FDA Oral History Interview, James Harvey Young with Alexander Schmidt, March 8–9, 1985, Chicago, Illinois, National Library of Medicine, Bethesda, MD., p. 51. Hereafter cited as Schmidt interview.

8. I have relied on transcribed segments from television news broadcasts obtained through the Vanderbilt Media Archives for much of the media analysis

in this chapter. Television provided a shared moment in the day for most families and one can assume that more families shared more even access and uniform exposure to the evening news broadcasts than any other media of the period. Exposure to printed sources including magazines, newspapers, and books varied far more widely by region as well as by economic, social, and educational distinctions. Until relatively late in the twentieth century, there was a choice of only three broadcast news stations, ABC, NBC, and CBS, and many top stories would appear on all three channels assuring a more uniform exposure to any given story. CBS's "Douglas Edwards and the News" was the first network news broadcast beginning in 1948. Soon NBC launched its "Camel News Caravan," hosted by John Cameron Swayze. News divisions lost money in the early years but the networks were eager to satisfy licensing renewal requirements that they provide "public service," and news broadcasts provided evidence of compliance with that mandate. In 1963, broadcasts expanded from 15 to 30 minutes. Videotape clips from the Vanderbilt Media Archives are available beginning around this time.

9. Suzanne White Junod and Lara Marks, "Women's Trials."

10. In the end, Commissioner George Larrick is reported to have given final clearance because he felt that young families might benefit from the ability to plan and space their families. Personal reminiscence from former FDA historian, Wallace Janssen, friend of and speechwriter for Commissioner Larrick. This was a commonly held view in family planning circles as well. See James W. Reed, *The Birth Control Movement and American Society: From Private Vice to Public Virtue* (Princeton, NJ: Princeton University Press, 1978, 1983).

11. W. M. Jordan to the editor, *Lancet* (18 November 1961): 1146–1147.

12. Trent Stephens and Rock Brynner, *Dark Remedy: The Impact of Thalidomide and its Revival as a Vital Medicine* (Cambridge: Perseus, 2001); Richard E. McFadyen, "Thalidomide in America: A Brush With Tragedy," *Clio Medica* 11:2 (1976): 79–93; Arthur Daemmich, *Pharmacopolitics: The Regulation of Drugs in the U.S. and Germany* (Chapel Hill: University of North Carolina Press, 2004).

13. NDA 10–976, Memo to Commissioner of FDA from Heino Trees, Division of New Drugs, 29 November 1962, RG 88, National Archives.

14. The committee itself was convened quietly. Even FDA's annual reports for 1961 and 1962 do not make mention of it or reported problems with the oral contraceptives.

15. "Contraindications to Oral Contraception," *British Medical Journal* 2(August 4, 1962): 315.

16. FDA Press Release, September 20, 1963, FDA History Office. William Grigg, "Pill for Birth Control Probed in 7 Deaths," Washington *Star,* August 4, 1962: A1; Gerald Grant, "Birth Curb Pill in Wide Use Here," Washington *Post,* August 5, 1962; John A. Osmundsen, "British Journal Issues Warning on an Oral Contraceptive Pill, New York *Times,* August 4, 1962.

17. Ibid.

18. Daniel Seigel and Phil Corfman, "Epidemiological Problems Associated with Studies of the Safety of the Oral Contraceptives," *Journal of the American Medical Association* 23 (March 11, 1968): 950–954.

19. Final Report on Enovid by the Ad Hoc Committee for the Evaluation of a Possible Etiological Relation with Thromboembolic Conditions, FDA, Department of Health, Education and Welfare, September 12, 1963, typescript in FDA History Office, p. 15; FDA Press Release, August 4, 1963, FDA History Office.

20. Sheila Jasanoff has described the government-wide establishment of such permanent advisory committees as the "fifth branch" of government. FDA was one of the first to employ them to extend the agency's own expertise. Sheila Jasanoff, *The Fifth Branch: Science Advisers as Policymakers* (Cambridge: Harvard University Press, 1990): 153.

21. Phillip Sartwell, Professor of Epidemiology, School of Hygiene and Public Health, Johns Hopkins University, was present as an advisor. His published studies, supported by FDA and the Advisory Committee, substantiated the risks first discovered by British researchers. P. E. Sartwell, T. Masi, et al., "Thromboembolism and Oral Contraceptives: An Epidemiologic Case Control Study," *American Journal of Epidemiology* 90:5 (1969): 365–80.

22. FDA Press Release, November 24, 1965; Advisory Committee on Obstetrics and Gynecology, Food and Drug Administration, *First Report on the Oral Contraceptives,* GPO, August 1966.

23. Watkins, *On the Pill*, 82. Advisory Committee on Obstetrics and Gynecology, Food and Drug Administration, *Second Report on the Oral Contraceptives,* GPO, August 1, 1969. In 1968, major large-scale prospective studies were begun, one by the Royal College of General Practitioners' in Britain, and the other by the Kaiser Permanente Medical Center in Walnut Creek, California. The Royal College's Oral Contraception Study was begun in 1966 with recruitment of users and controls in 1968. 1400 general practitioners in the United Kingdom recruited over 23,000 OC users and 23,000 matched controls that had never used oral contraceptives for the study. The first major conclusions were published in 1974, and as late as 1984, there were still 19,000 women under observation. Royal College of General Practitioners, *Oral Contraceptives and Health: An Interim Report from the Oral Contraception Study of the Royal College of General Practitioners* (New York: Pitman, 1974). The Walnut Creek Study also began in 1968 with 21,000 women admitted to the study and then divided into "takers," "past takers," and "non-users" of oral contraceptives for the purposes of analysis. Results from these studies, studying the association between the oral contraceptives and breast cancer, began to appear in 1972, but the main body of the report was published in 1975. The Walnut Creek Contraceptive Study, *A Prospective Study of the Side Effects of Oral Contraceptives, Vol. 1: Findings in Oral Contraceptive Users and Non-Users at Entry into the Study* (Washington D.C.: GPO, 1975).

24. James Harvey Young, *Pure Food: Securing the Federal Food and Drugs Act of 1906* (Princeton: Princeton University Press, 1989).

25. Richard A. Merrill and Peter Barton Hutt, *Food and Drug Law: Cases and Materials* (New York: Foundation Press, 1980): 134–166. Charles O. Jackson, *Food and Drug Legislation in the New Deal* (Princeton: Princeton University Press, 1970).

26. Michelle Meadows, "CDER: Promoting Safe and Effective Drug Use," *FDA Consumer,* January/February, 2005.

27. FDA Annual Reports, 1966, in *Food and Drug Administration Annual Reports, 1950–74* (Washington D.C., 1976): 197. Hereafter referred to as *Annual Reports.*

Federal Register 32:177 (Wednesday, September 13, 1967): 13008.

29. Ibid.

30. FDA Press Release, April 29, 1968, FDA History Office; W. H. W. Inman and M. P. Vessey, *British Medical Journal* 2 (1968): 193–199; M. P. Vessey and R. Doll, "Investigation of Relation Between Use of Oral Contraceptives and Thromboembolic Disease," *British Medical Journal* 2 (1968): 199–205. These studies found a seven to tenfold increase in morbidity and mortality due to thromboembolic diseases in women taking oral contraceptives. The controlled, retrospective studies involved 36 reported deaths and 58 hospitalizations due to "idiopathic" thromboembolism. They concluded that the statistical differences between users and nonusers of the Pill were "highly significant."

31. "Dear Doctor" letter, June 28, 1968, James L. Goddard, M.D., Commissioner of FDA, FDA History Office files.

32. FDA Internal report, 7/24/68, Robert M. Hodges, M.D., Associate Director, Office of New Drugs, Bureau of Medicine, FDA History Office. In this report, Hodges acknowledged that they had affirmed that thromboembolic problems were evident in patients on both combination and sequential oral contraceptives and that when brands were shifted, a second event had occurred in some patients. "The conclusion at the date of writing is that these products, while they have to be used with due regard to the patient's background are safe and effective and are marketed for the highly important purpose of population control."

33. Barbara Seaman, an early critic of Enovid, claimed that the oral contraceptive enjoyed journalistic "immunity" in its early days. Personal communication with the author.

34. Drew Pearson and Jack Anderson, "FDA Lags on Data for 'Pill' Fatalities," Washington *Post,* March 19, 1969.

35. Memorandum, undated, to Secretary of Health, Education and Welfare, from FDA Administrator through Assistant Secretary for Health and Scientific Affairs, Consumer Protection and Environmental Health Service, FDA History Office files.

36. Ibid.

37. *Second Report on the Oral Contraceptives,* p. 6.

38. Ibid., 3.

39. Watkins, *On the Pill;* Andrea Tone, *Devices and Desires;* Lara Marks, *Sexual Chemistry.*

40. Watkins, *On the Pill,* 115.

41. CBS, Tuesday, June 23, 1970, "AMA/Pill Warning." The most detailed discussion of the controversy over the adoption of the PPI is found in Watkins, *On the Pill,* 120–128.

42. So many drafts of the proposed "patient package insert" were circulating in the first six months of 1970 that Deputy Assistant Secretary for Population Affairs requested a chronology of the OC labeling for patients. (John Jennings, M.D., Assistant to the Commissioner for Medical Affairs to Louis M. Hellman, M.D., Deputy Assistant Secretary for Population Affairs, Department of Health, Education, and Welfare, July 13, 1970, FDA History Office files.) On March 4, 1970, one version had been presented to the Nelson Committee and had been published in newspapers around the country. Dr. Hellman's own version had been published in the Washington Drug Letter on March 23, 1970 and entered into the minutes of the Ob-Gyn Advisory Committee's 15th meeting on April 1, 1970. The proposed labeling was published in the *Federal Register* on April 10, 1970. Commissioner Edwards offered his own assessment in June 1970, which would become the basis of the final ruling, effective July 11, 1970. "Statement of Policy Concerning Oral Contraceptives Labeling Directed to Users" (CFR 130.45), 35:113, June 11, 1970. See Statement by Charles C. Edwards, M.D., Commissioner of Food and Drugs, before the Subcommittee on Intergovernmental Relations, Committee on Government Operations, June 9, 1970, FDA History Office files. Also, HEW News Release, June 9, 1970, FDA History Office files.

43. Taking into account that the Pill was composed of potent steroid hormones "that affect many organ systems," the fact that they were used, often without adequate medical supervision, for a variety of conditions, by healthy women and for long periods of time, the Administrator had decided that "they represent, therefore, the prototype of drugs for which well-founded patient information is desirable. . . . The Commissioner of Food and Drugs is aware that this represents a departure from the traditional approach to the dissemination of information regarding prescription drugs via the doctor/patient relationship, and stresses that it is not intended to weaken or replace that channel, but rather because of the unusual pattern of use by these drugs, to reinforce the efforts of the physician to inform the patient in a balanced fashion of the risks attendant upon the use of oral contraceptives." "Proposed Statement of Policy Concerning Oral Contraceptive Labeling Directed to Laymen," *Federal Register* 35: 70 (April 10, 1970).

44. "Statement of Policy Concerning Oral Contraceptives Labeling Directed to Users" (CFR 130.45), *Federal Register* 35: 113 (June 11, 1970): 9002. For lessons learned and early assessments of the first patient package insert, see L. Flecken-

stein, J. Pieter, et al., "Oral Contraceptive Patient Information: A Questionnaire Study of Attitudes, Knowledge, and Preferred Information Sources," *Journal of the American Medical Association* 235:1331–1336 (1976): 84–89.

45. NBC News, Monday, March 30, 1970, "Finch/Pill"; CBS, Monday, March 9, 1970, "Feminist Movement/Now Protest"; ABC, Wednesday, March 4, 1970, "FDA/Pill."

46. Nea D'Amelio, "How 334 Gynecologists View the Pill," *Medical Times* 98 (October 1980): 206–12. Cited in Watkins, *On the Pill.*

47. Committee on the Safety of Drugs, "Combined Oral Contraceptives: A Statement," *British Medical Journal* 2:5703 (April 25, 1970): 231–2.

48. Royal College of General Practitioners, *Oral Contraceptives and Health: An Interim Report of the Royal College of General Practioners* (London: Pitman, 1974).

49. Roger M. Rodwin, "The Freedom of Information Act: Public Probing into Private Production, *Food, Drug, and Cosmetic Law Journal* 28 (August, 1973): 533–544. Jeremy Lewis, "The Freedom of Information Act: From Pressure to Policy Implementation," unpublished dissertation, Johns Hopkins University, 1982. Freedom of Information Act Amendments, P.L. 93–502, 88 Stat. 1564 (1974).

50. Ibid.

51. "The Pill," *FDA Consumer* 6:10 (December 1972):16–17.

52. Merrill Interview, p. 22–23. Schmidt Interview, p. 88. Wayne Pines, editor of *FDA Consumer*, quotes Charles Edwards as saying "Throughout the FDA it is our daily business to recognize and to respond to higher consumer values—and where we can and should, to shape those values through communication as well as regulation." Wayne Pines, "FDA and The Age of the Consumer," *FDA Consumer* 6:7 (September 1972): 26.

53. Ibid., 90. Schmidt gave himself a C/C plus in his dealings with the press. "I viewed dealing with the press not as important as managing the agency . . . I know now that dealing with the press was far more important than I thought early in my administration. It is a bully pulpit and I didn't use it enough."

54. CBS, Thursday, October 20, 1973 "FDA/Birth Control Pill"; ABC, Thursday, September 20, 1973, "FDA/Birth Control Pills"; NBC, Thursday, September 20, 1973, "FDA/Birth Control Pills."

55. "Oral Contraceptive Patient Labeling, Proposed Venereal Disease Warning Statement," *Federal Register* 39: 77 (April 19, 1974): 13972.

56. Ibid. A former FDA physician in practice at the time the Pill was introduced admitted that he and his colleagues did what would now be considered a lot of "unnecessary" tests on his early Pill patients until he was convinced that they were not fundamentally changing any critical reproductive or circulatory systems.

57. W. H. W. Inman and M. P. Vessey, *British Medical Journal* 2 (1968): 193–199; M. P. Vessey and R. Doll, "Investigation of Relation Between Use of Oral Contraceptives and Thromboembolic Disease," *British Medical Journal* 2:599 (1968): 199–205.

58. Clifford Kay, A. Smith, B. Richards, "Smoking Habits of Oral Contraceptive Users," *Lancet* 2:7632 (December 6, 1969): 1228–1229.

59. M. F. Oliver, "Oral Contraceptives and Myocardial Infarction," *British Medical Journal* 2:703 (April 25, 1970): 210–213.

60. H. Fredericksen, and R. T. Ravenhold,"Oral Contraceptives and Thromboembolic Disease," *British Medical Journal* 4:633 (December 21, 1968): 770; H. Fredericksen, R. T. Ravenhold, "Thromboembolism, Oral Contraceptives, and Cigarettes," *Public Health Reports* 85 (March 1970): 197–205. Another correspondent even suggested a mechanism of action, explaining that "clot retraction" had been identified among smokers. In the case of a thromboembolism, clot retraction could delay reestablishment of circulation and aggravate the effects of a thrombus. F. Nour-Elden, "Oral Contraceptives and Thromboembolic Disease," *British Medical Journal* 4:674 (October 4, 1969): 51.

61. "A Statement of the Committee on Safety of Drugs: Combined Oral Contraceptives," *British Medical Journal* 2 (April 25, 1970): 231–232.

62. ABC, Monday, February 17, 1975, "Stroke/BirthControl Pills"; CBS, Monday, February 17, 1975, "Birth Control Pills."

63. CBS, Friday, May 9, 1975, "Birth Control."

64. *Population Reports,* "Oral Contraceptives," A:2 (March 1975): A-29.

65. J. I. Mann, M. P. Vassey, M. Thorogood, and R. Doll, "Myocardial Infarction in Young Women with Special Reference to Oral Contraceptive Practice," *British Medical Journal* 2:5965 (May 3, 1975): 241–245. This study compared the frequency of use of oral contraceptives in women discharged from the hospital with a MCI diagnosis with the frequency in age-matched controls randomly selected from the same hospital. Age 30–39, the assessed risk was 2.7 (5.6 among users and 2.1 among nonusers), but for women 40–44, the relative risk was assessed at 5.7 (56.9 among users and 9.9 among nonusers). J. I. Mann, W. H. W. Inman, "Oral Contraceptives and Death from Myocardial Infarction," *British Medical Journal* 2 (May 3, 1975): 245–248. This second study compared the frequency of use of OCs in patients who had suffered a fatal MI with the frequency of use in a control population. The relative risk was 2.8 in women aged 30–39 (5.4 incident in OC users; 1.9 in nonusers) and 4.7 (54.7 incident in OC users; 11.7 in nonusers) for those 40–44.

66. CBS, Wednesday, August 27, 1975, "Birth Control"; ABC, Wednesday, August 27, 1975, "Birth Control."

67. NBC, Wednesday, August 27, 1975, "Birth Control." The next day, only NBC reported that the American College of Obstetrics and Gynecology and Planned Parenthood had responded that much more research was needed on the subject of women over 40 taking the Pill. NBC, Thursday, August 28, 1975, "Birth Control."

68. FDA Drug Bulletin, "Risk of Myocardial Infarction in Users of Oral Contraceptives," July/August 1975, p. 10.

69. Ibid.

70. Ibid.

71. Thomas J. Glynn, and Joseph W. Cullen, Division of Cancer Prevention and Control, NCI, NIH, "Smoking and Women's Health," *Women's Health: Report of the PHS Task Force on Women's Health Issues, Vol. II,* DHHS Publication No. 88–50206, October 1987: 86–91. Table reports decrease in male smoking rates from 52.5% in 1935 to 36.9% in 1979, but a concomitant increase in female smoking rates from 18.1% in 1935 to 28.2% in 1979, and concludes that 20% of teenagers are regular smokers. "In women who use ocs and also smoke, the risk of heart attack is increased approximately tenfold over that in women who neither smoke nor use oral contraceptives."

72. CBS, Thursday, October 16, 1975, "Birth Control Pills"; FDA Talk Paper, "Birth Control Pill Labeling," October 16, 1975, FDA History Office files.

73. ABC, Thursday, October 16, 1975, "Birth Control Pills."

74. On February 25, 1976, CBS and ABC reported that drug manufacturers would discontinue making three so-called sequential birth control pills (Oracon, Ortho-Novum SQ, and Norquen) because FDA had come to consider "combination" pills to be safer. Combination pills contained both estrogen and progesterone in each pill, whereas the sequentials provided estrogen first, and then progesterone. "Unopposed" estrogen was a major concern of researchers beginning to look at the links between breast, uterine cancer, and the oral contraceptives. CBS, Wednesday, February 25, 1976, "Birth Control Pills"; ABC, Wednesday, February 25, 1976, "Birth Control Pills"; HEW News Release, February 26, 1976, FDA History Office files.

75. CBS, Thursday, June 3, 1976, "Special Report/Part 3/Birth Control Pills."

76. "Oral Contraceptive Drug Products: Physician and Patient Labeling," *Federal Register* 41: 236 (December 7, 1976): 53633–42; "Informing Women about 'The Pill,'" *FDA Consumer* 11:1 (February 1977): 21. "Oral Contraceptive Drug Product: Revision of Physician and Patient Labeling," *Federal Register* 42:103 (May 27, 1977): 27303–4. The British response came from the Committee on the Safety of Medicine and was published in the *British Medical Journal* 2:6092 (October 8, 1977): 965. "While the numbers in the studies are too small to give precise conclusions concerning overall risk, the trend noted earlier that the risk of arterial thrombosis associated with pill use increases with age is supported. This risk is aggravated by cigarette smoking. Because of changes in the composition of the pill, which has occurred over the time of this study, it is impossible to make an assessment based on current products. Based on this result, changes in the warning labels are unnecessary except to emphasize the increased risk for women in the later age group, especially those who are cigarette smokers.

77. HEW News Release, December 5, 1976, P76–36, FDA History Office files.

78. CBS, Tuesday, June 1, 1976, "Special Report/Part 1/Birth Control Pills."

79. CBS, Wednesday, June 2, 1976, "Special Report/Part 2/Birth Control Pills."

80. CBS, Thursday, June 3, 1976, "Special Report/Part 3/Birth Control Pills."

81. Ibid.

82. NBC, Wednesday, January 11, 1978, "HEW Smoking."

83. ABC, Tuesday, April 10, 1979, "Closeup/Birth Control: Risk Benefit Factor"; NBC, Wednesday, January 11, 1978, "HEW Smoking." Bold as Califano's attack on smoking was, Jennings announced with some sly wit that the Secretary of HEW had not gone so far as to butt heads with his colleague, the Secretary of Agriculture, by criticizing tobacco subsidies to farmers. When asked why not, Califano reportedly replied, "I'm not going to tilt at that windmill." [President Carter was a southern farmer.]

84. "Final Rule: New Drugs Requirement for Labeling Directed to the Patient," *Federal Register* 43: 21 (January 3, 1978): 4214–23.

85. ABC, January 24, 1978, "Birth Control Warnings."

86. CBS, January 24, 1978, "Birth Control/Smoking"; cigarette smoking increases the risk of serious adverse effects on the heart and blood vessels from oral contraceptive use. This risk increases with age and with heavy smoking (15 or more cigarettes a day) and is quite marked in women over 35 years of age. Women who use oral contraceptives should not smoke.

87. Ibid.

88. More coherent explanations of the risks began appearing in the published literature as well. See D. Stone, S. Shapiro, L. Rosenberg, et al., "Relation of Cigarette Smoking to Myocardial Infarction in Young Women," *New England Journal of Medicine* 298 (1978): 1273–6. "The risk of MI and stroke that is attributable to oral contraceptives is concentrated among women with other risk factors, in addition to being concentrated among women 35 or older. However, cigarette smoking is far more prevalent among women of reproductive age in Britain and the U.S. than are other risk factors for mci and stroke." See also, Bruce Stadel, "Medical Progress, Oral Contraceptives and Cardiovascular Disease," *New England Journal of Medicine* 305 (September 10, 1981): 612–18; (September 17, 1981) 672–7. "In general, oral contraceptives have been found to multiply the effects of age and other risk factors for myocardial infarction and stroke rather than add to them. Therefore, the risk of myocardial infarction and stroke that is attributable to oral contraceptives is primarily concentrated among older women and women with other risk factors."

89. Ortho-Evra "Patient/Doctor": 60, submitted for review 3/28/05, FDA #03E180, Division of Drug Marketing, Advertising, and Communications, Center for Drug Evaluation and Research, FDA. Thanks to Melissa Moncavage in that office for her assistance.

90. C. R. Kay, "The Royal College of General Practitioners' Oral Contraception Study: Some Recent Observations," *Clinical Obstetrics and Gynaecology* 11:3 (December 1984): 759–786.

91. E. Ketting, "Contraceptive Needs of Women Over 35," *Maturitas*, supplement 1 (1988): 23–38.

92. *Federal Register* 52: 76 (April 21, 1987): 13107–9.

93. *Federal Register* 54: 100 (May 25, 1989): 22585–88.

94. J. Newton, "Contraception for the Woman Aged 35 and Over," *Maturitas*, supplement 1 (1988): 89–97. Sharon Snider, "The Pill: 30 Years of Safety Concerns," *FDA Consumer* (December 1990): 10.

95. C. R. Kay, "Some Recent Observations," 786.

96. Press release, Virginia Commonwealth University, Science Daily, "Low-Dose Oral Contraceptives May Increase Risk for Heart Attack or Stroke," July 7, 2005.

97. Lorine Goodwin, *The Pure Food, Drink, and Drug Crusaders, 1879–1914* (NC: McFarland & Co., 1999); Gwen Kay, *Dying to Be Beautiful: The Fight for Safe Cosmetics* (Columbus: Ohio State University Press, 2005).

5

STIMULANTS

Not Just Naughty
50 Years of Stimulant Drug Advertising

Ilina Singh

(Re)constructing a history of methylphenidate (Ritalin)

How do we learn about a drug's history? What resources are available to us when we begin to wonder where a particular drug came from, how it entered the marketplace, the clinic, and the domestic realm? One might follow the drug (or its ancestors) through the clinical literature, exploring its emergence on the scene, who was using it, what for, and with what effects. One might write a chemical history, linking the creation of the drug on a molecular level to other drugs in its molecular field, and the men (usually) and the laboratories whose succor brought the drug into being.

It is extraordinary that to date we are lacking in even these more straightforward historical accounts of the stimulant drugs used so widely in contemporary psychiatry.[1] This is not because these drugs are esoteric; rather, they are household names, icons for contemporary Western socio-cultural preoccupations: Ritalin, Concerta, Dexedrine, Adderall. We have a few histories of another group of popular psychiatric drugs, the antide-pressants and the minor tranquilizers, but most people still believe that the most famous of these drugs, Ritalin (methylphenidate), is a "new" drug that came onto the market some time in the 1980s along with concerns about hyperactivity in children. As this chapter explains, Ritalin is not a new drug; it first came on the U.S. market in 1957. But experiments involving stimulant drugs and problematic behavior in children have been on-going since at least the 1920s. What is new is the extraordinary world-wide growth in medical use of methylphenidate, predominantly for symptoms of Attention Deficit/Hyperactivity Disorder (ADHD). Consumption of methylphenidate in the United States outpaces all other countries; between 1991 and 1999, domestic sales of the drug had grown by 500%. Significant growth in consumption in this period is also evident in Canada, New Zealand, Australia, and Norway; however, levels of consumption in these countries are still a fraction of U.S. levels. [2] Approximately 85% of the world's methylphenidate is currently consumed in America.[3] U.S. epidemiological estimates for ADHD are problematic and vary widely, from 1% to 10% of the school-age population. By contrast, estimates in other Western countries such as the United Kingdom vary from under 1% to 3% of the school-age population, and in non-Western countries ADHD and Ritalin are only just being discovered.

Clearly, then, there is an important socio-cultural component to the growth of the ADHD diagnosis, Ritalin, and other stimulant drugs. This chapter contributes to reconstructing the socio-cultural story of the stimulant drugs in America, as part of a larger project which explores how stimulant drugs came to be persuasive tools for behavior management and modification in young children. As such, the focus is on methods of persuasion and modes of representation of these drugs and the symptoms they treat. There is an embedded interest in the major actors—both in those who are doing the persuading and in those who are being persuaded: Doctors, of course, occupy both these actor positions, more or less simultaneously. However, other actors in this story have remained shadowy figures, particularly one of the main actors, the pharmaceutical industry. Why so many shadows in this history? A major problem is that the pharmaceutical industry, specifically Novartis (formerly Ciba) has been only marginally helpful to scholars attempting to piece together the history of the most famous stimulant drug, Ritalin.[4] The lack of industry transparency in relation to drug development, testing, and promotion is endemic; moreover, the sustained critique of Ritalin is likely to support a tendency to be closeted within Novartis. In the existing historical treatments of hyperactivity, the pharmaceutical industry has been villainized for its contribution to the medicalization of children's problem behaviors via various manipulative schemes.[5] But such arguments have necessarily relied largely on anonymous industry informants and anecdotal evidence to substantiate their claims.

The Problem of Evidence

Clearly there is a problem of evidence, or documents, for those wanting to write a history of stimulant drugs. While it is difficult to access the pharmaceutical companies behind these drugs, it is possible to access their public documents. Stimulant drug advertisements are public documents, created and sanctioned by the pharmaceutical company. Treating these advertisements as documents contributes to the history of stimulant drugs in the most basic sense that it provides a new archive to access for scholars, who want to deepen and broaden the existing historical understanding of these drugs. Drug advertisements are an important part of a drug company's persuasive arsenal; they are used to educate the doctors and the public about drugs, and they are used to shape attitudes about and awareness of the body, health, and well-being as part of a campaign to create desire for pharmaceutical products. Advertising drugs to doctors may be an effective form of persuasion and marketing. Ads in clinical journals have long been an important source of drug information for doctors; a 1974 FDA survey reported that 50% of doctors reported using journal advertisements for self-educational purposes.[6] More recent studies have shown that doctors rely on commercial sources of drug information, especially if they had been practicing more than 15 years,[7] and that doctors are more likely to write prescriptions for drugs they have seen advertised.[8–10] Direct-to-Consumer drug advertising, which is allowed in the United States and New Zealand but in no other country in the world, is also proving to be an effective means of drug promotion. Doctors are more likely to prescribe drugs when consumers ask for drugs by name, and consumers ask for products they have seen advertised.[11]

It must also be said, however, that it has been difficult to establish a firm causal link between drug industry promotional activities and doctors' prescribing decisions. The most obvious association found is that doctors who report relying more on promotional material prescribe more drugs, prescribe less rationally, and prescribe newer drugs earlier than other doctors However, other characteristics of doctors, such as age, academic level and involvement, and clinical experience may be important factors in this association.[12–14] In what follows, it should be remembered that advertisements in journal articles are part of a larger drug promotion campaign, which included, for Ritalin, educational videos for clinicians, drug representative visits to clinical and public sites, and funding research into child hyperactivity and drug response. Therefore, these advertisements are most usefully viewed as additional historical resources which can help deepen analyses that include work with a variety of relevant documents. Advertisements can be viewed as discursive spaces, the meeting point of a variety of prominent cultural discourses and rhetorical strategies. In particular, these advertisements point to the clinical-cultural tropes the pharmaceutical industry has both (re)constructed and deployed to persuade first the clinic, then the lay public, of the need for these drugs.

There are two overlapping aims in this chapter. The first is simply to present some of the advertisements for stimulant drugs chronologically, from the late 1950s until the present. The visuals are allowed to speak for themselves to a certain extent, alongside a historical and interpretive context against which to "read" the ads and flesh out the history of these drugs. In the course of this chronology, however, there is a second aim: To make the argument that the contemporary phenomenon of rising ADHD diagnoses and stimulant drug use may be explained, in part, by an older association between problem boys and their problematic mothers. Put even more simply, mothers matter to this phenomenon, but how and why they matter cannot be captured by simplistic and biased mother/parent-blaming narratives. This chapter explores the ways in which stimulant drug advertisements capitalize on the association between problem boys and problematic mothers in order to sell disease and drugs to psychiatrists, and later to sell a drug-assisted relationship and lifestyle to mothers.[15]

Ritalin: Gentle, Collaborative, Catholic

Ritalin was first marketed in 1955 by Ciba Company for narcolepsy. Early advertisements show tired patients, whose fatigue is associated with a wide range of psychiatric disorders, including chronic fatigue, depression, and dementia. The text and visuals of Ritalin advertisements throughout the 1950s and 1960s suggest an effort to position Ritalin as a drug that in a sense defies diagnosis: It is useful in the treatment of most psychiatric diagnoses. It is also amenable to working alongside other treatments, including more powerful drug treatments and psychotherapy. Two groups of potential Ritalin users are depicted: the middle-aged and the elderly. Within these groups, Ritalin candidates are male or female, and White. Initially, Ritalin seems to have been marketed more directly for geriatric patients; ads highlight the benefits of Ritalin's gentle action and mild side effects for this vulnerable group. Toward the end of the 1960s, however, adverts appear that specifically expand the range of candidates for Ritalin. One such advertisement shows the profiled faces of four men, ranging in age from approximately 40 to 80. All of the men have similarly sad expressions on their faces, although their psychiatric diagnoses are different. The text of the advert reads, "All are candidates for Ritalin." The symptoms depicted in these advertisements are in no way frightening or flagrant; rather, they are familiar depictions of sadness. When pictured alone these men and women look lonely and depressed; they gaze into the distance, rarely meeting the eye of the viewer.[16]

While doctors are infrequently pictured in the adverts, they are, of course, suggested in the gaze of the viewer, and indeed, in one advertisement, the reader sees the (female) patient through the eye glasses and spectroscope of the pictured (male) doctor, creating a direct visual bond between the reader-doctor and the imaged doctor. The most prominent visual in this advert is the doctor's fist which both holds the spectroscope and serves as a forceful barrier between himself, the reader-doctor, and the female patient. Indeed, there is a great deal of authority and power suggested in the tools of the trade (spectroscope) and the doctor's fist, which both frame the patient and hold/push her back. Having established the authority of the clinician visually, the text of this advert can exhort the clinician to act gently, not to move too quickly to powerful drugs to treat vulnerable middle-class women, but rather to try "a gentle stimulant with few side effects" and to "reserve more potent agents for more serious conditions." Such advertisements serve to position both Ritalin and the prescribing clinician as gentle, sympathetic entities (while visually affirming medical authority); just as the patient and her symptoms are familiar, so too is the doctor familiar and his practices thoughtful and unthreatening.

Such images can be linked to a project of domestication of psychiatric authority, as represented by doctors, diagnosis, and treatments. While these advertisements encourage the use of drug therapies that have been more traditionally viewed as part of a biological psychiatry, the visuals and text of these advertisements suggest a not-so-subtle interaction with psychoanalytic approaches to "every day" problems of living. This will be discussed further in a later section of this chapter; at this point, it is sufficient to point out that some of the advertising for Ritalin consciously employs a psychoanalytic trope in its ads, binding Ritalin to a sympathetic male authority that offers a safe haven, and a safe relationship, for a group of middle-class females—the "worried well."

Should psychoanalytic psychiatrists still have been worried about using psychotropic remedies in patient therapy, there was another set of adverts to speak directly to the collaborative qualities of mild stimulants. Two of these ads appeared in 1958, soon after Ritalin first came on the market, and suggest that drug companies were aware at this point that psychoanalytic models were sufficiently in ascendance during this time, and that marketing drugs specifically as adjuncts to psychotherapy could be efficacious. An advertisement for Dexedrine (which was earlier indicated for compulsive overeating) shows Rorschach-like images and cites a published clinical finding that "Dexedrine may function impressively as a specific adjuvant in psychotherapy." The ad for Ritalin is an attempt to simultaneously market a novel form of the drug and position Ritalin as a useful adjunct to therapy. The novel form of Ritalin is "new parenteral Ritalin," a Ritalin injection. The byline reads, "Help psychiatric patients talk." The ad is largely text, interspersed with images of a male doctor and a male patient; the latter is obviously finding talking effortful at first. Parenteral Ritalin offered doctors the opportunity to make their patients more "verbal" and "cooperative" quickly—the text promised "rapid" action . . . "in as little as five minutes." (One can only imagine what a shot of amphetamine would do to a patient in the doctor's office; this form of Ritalin appears to have been taken off the market very quickly, for reasons about which we can only speculate. For one, abuse potential would be high. By doctors or patients or both?)

While early ads for Ritalin and Dexedrine showed adults exclusively, the use of psychotropic drugs for behavior problems in children was clinically well established by the time Ritalin came on the market in 1955. Results of experiments with the stimulant drug Benzedrine were published by Charles Bradley in 1937;[17] in the following decade, subsequent articles by Bradley and others provide evidence of increasing clinical experimentation and refinement.[18–20] However, the most commonly used drug treatment for children with behavior problems was still the minor tranquilizers. Tranquilizers for use with children were marketed to clinicians in the *American Journal of Psychiatry* in the late 1950s and early 1960s. "Calmative Nostyn" (1958) was declared "safe for your little patients too" and indicated for use in "hyperactive" as well as "emotionally unstable" children. Prozine (1959) controlled "acute behavior problems" in children. An Atarax campaign (1962) declares that this is "a widely prescribed tranquilizer in pediatrics." The benefits of tranquilizing children to improve school performance is indicated in many of these ads; Atarax, for example, "promptly relieves overt anxiety, lengthening the child's attention span for better schoolwork and easing his relations with teachers, classmates, and parents."

Constructing the "Problem" Child in Medical Advertisements for Stimulant Drugs

These ads give us several important clues to the clinical landscape in relation to psychotropic drug treatments for children. Anxious children, hyperactive children, and emotionally disturbed children were well established clinical entities by the late 1950s, such that drugs could be marketed directly for use for these diagnoses. The focus on "the hyperactive" child in the late 1950s may reflect growing clinical awareness and definition of hyperactivity as a disorder of childhood.[21] Emotional disturbance had long been an ambiguous clinical description of a child with a wide variety of behavioral symptoms whose etiology was largely unknown. Symptoms were thought to be secondary to an underlying primary disorder, and treatment of the secondary symptoms was considered vital to the effort of containing this underlying disorder.[22]

Why Boys?

Another significant clue to contemporary and clinical perspectives on the use of these drug treatments in children is the fact that, between 1955 and 1998, all children pictured in advertising campaigns for minor tranquilizers and stimulant drugs are boys. If the persuasiveness of advertisements depends in part on the advertiser's ability to draw upon familiar cultural tropes and patterns of understanding, then this exclusive, and persistent, focus on boys must be seen as meaningful, not incidental. Drug campaigns both rely upon and construct gender ideologies and gendered behaviors.[23–25] We need to interrogate the gendering of pathological behavior in children in these advertisements, because by doing so we may come to understand something about the remarkable gender skew in psychiatric diagnosis more generally. In the case of stimulant drug advertising for behavior problems in children, we need to ask the question: Why boys?

What is the basis for the series of equations presented by these ads: children who exhibit behavior problems are boys; behavior problems in boys fall under medical purview; psychotropic drugs are appropriate tools for controlling behavior in boys.

These questions are meant to be part motivational rhetoric; it is impossible to do justice to them in a single chapter. We have little substantive understanding of why an increasing number of school-age boys in America, and around the world, are taking stimulant drugs. We have a number of broadly justified analyses that point fingers at a combination of cultural, social, clinical, familial, biological, and genetic factors. But too often these analyses do not provide an understanding of the *sustained* gendering of the school-age population of stimulant drug candidates. In Conrad & Schneider,[26] for example, we had a compelling argument about ADHD as the medicalization of delinquent and deviant behavior. While delinquency is also a gendered phenomenon, this argument did not develop a gender analysis. More recently, a new generation of experts in boys' psychology have argued that we are "medicalizing boyhood" with stimulant drugs in a contemporary setting which is so fast paced and competitive that boys are forced to live up to a "culture of masculinity" much before their time.[27–28] This argument can at times assume an almost idyllic past in which boyhood was imbued with a Rousseau-like innocence and freedom. And yet what we are learning from the developing history of stimulant drugs is that, even half a century ago, quite normal-seeming boys were presented as appropriate candidates for drug intervention. This appropriation of boyhood by medicine and the pharmaceutical industry is an area that requires exploration. So perhaps a useful way into an understanding of "why boys" is to explore the gendering of normality and abnormality in children's behavior, and to ask how this gendering is achieved.

Problem Boys and Problematic Mothers

How are behavioral "normality" and "abnormality" in boys represented in these visual clinical texts? To begin to answer this question, we need to locate the particular gendering of behavior in children represented in these adverts, within our earlier discussion about the subtle referencing of gendered relational dynamics inside the clinic. Because while an advert may only picture one boy, there are at least two other centrally important figures implied in the discursive space of the advertisement. One is, of course, the (male) clinician. The other is the boy's mother. Representations of the normal and the abnormal in young boys' behaviors are constructed in the context of this relational dynamic.

If one were to take a single volume of the *American Journal of Psychiatry* (*AJP*) as a whole in the late 1950s, remove all the articles and leave just the adverts, one might begin to view this collection of advertisements as a picture-book. Each page has its own individual visual and verbal narrative, with a particular message. However, each page also references the other pages in the book, such that there is a collective story built about clinical perspectives on mental illness and the characters that best represent the profession's ability to provide expert care and treatment. If this were a volume from the early 1960s, we would learn that middle-class, middle-aged White women were anxious, depressed, and worried. They visited White male psychiatrists who viewed them with fearful longing[29] and sent them home with tranquilizers. These women had children, but only boys, who misbehaved. The boys were hyperactive and unstable. If we added in some pages from a later *AJP* volume, in the 1970s, we would begin to get a sense of some narrative connections: The women's anxiety and depression exist alongside the boys' problem behaviors. Women accompany their sons to the doctor's office. Women look a great deal happier following treatment of their sons. We begin to get a sense that the condition of the mother relates to the condition of the son. Indeed, the picture-book is telling us that these conditions refer to each other in some way; the drugs advertised appear less to be treating individual conditions and more to be treating relationships and interactions.

Historically, biological child psychiatry, as viewed through the lens of stimulant drugs, contains a psychoanalytic assumption of an intimate relationship between a problem boy and his problematic mother. It is this cultural trope which in part supports the rise of stimulant drug use in children. Indeed, a degree of integration of biological psychiatry and psychoanalysis in treatment approaches to "the worried well" should now be unquestioned. The clinical literature in pediatrics and child psychology demonstrates this integration throughout the 1940s and 1950s,[30–31] and advertising of stimulant drugs to clinicians in the 1950s and 1960s directly emphasized the interconnections between drug treatments and psychoanalytic approaches. Moreover, a similar point has been illustrated persuasively, and more expansively, in Jonathan Michel Metzl's discussion of advertisements for the minor tranquilizers and antidepressants.[32] If Metzl's project is to illumine the "Freud in Prozac," mine has been, more narrowly, to explore the professional preoccupation with mother in the impulse to medicalize behavior problems in boys. One might say that the object here is to illumine the mother in Ritalin. The following analysis scrutinizes a set of 1970s *AJP* advertisements for Ritalin, in order to support the claim that this triad of mother-son-clinician is the relational ground upon which the understanding of normal/abnormal behavior in boys is constructed.

Ritalin as a Niche Drug for Minimal Brain Dysfunction (MBD)

Until 1971, advertisements for Ritalin in the *AJP* do not depict children. However, in 1960s advertisements, the text does provide dosage information for "children with functional behavior problems," suggesting that the clinical practice of giving Ritalin to children was reasonably well established during this period. Indeed, by the 1960s, several influential journal articles had documented the superior benefits of methylphenidate over tranquilizers in the treatment of children's overactive and distracted behaviors.[33–35] The group of children whose behaviors benefited most from methylphenidate treatment suffered from what the authors called "chronic brain syndromes."[36] Two years later, a variant of such syndromes, Minimal Brain Dysfunction (MBD), reemerged as an important childhood diagnosis through the publication of a U.S. Public Health Service Report. MBD presented Ciba with the opportunity to market Ritalin for children as a niche drug.

The 1966 U.S. Public Health Service publication, "Minimal Brain Dysfunction in Children: Terminology and Identification," by Samuel Clements, attempted to pull together a large amount of material on MBD and similar disorders, in order to present a definitive diagnostic guide. In Clements' review of the literature, he found that MBD had been described by at least 38 different terms, including hyperkinesis and hyperactivity syndrome. Thirty years' worth of clinical evidence suggested that methylphenidate was an effective treatment for this particular cluster of behaviors, and given the ambiguous nature of the diagnosis, it would be difficult not to begin to use the drug as a pharmacological scalpel. Public health officials sought to educate doctors as well as the public about the nature of MBD, distributing booklets, pamphlets, and short films on how to recognize, evaluate, and treat MBD. The popular press cast the newly hot diagnosis into its ever fiery debate on the growth of the drug industry and medical diagnosis, the use of psychotropic drugs in everyday life, and the use of psychotropic drugs in children.[37]

It is against this background that Ritalin, the gentle, collaborative drug originally marketed for a tired, elderly population, becomes fundamentally associated with a population of hyperactive youth. The connection between Ritalin and MBD (or hyperactivity) can be seen as part of a project of "branding" a condition, that is, elevating its importance, creating consensus about the condition, and deciding on the best form of treatment.[38] However, even as a niche drug, Ritalin had to retain its catholic heart—MBD was after all an umbrella term for at least 38 other diagnostic terms. Its gentle persona served it well in this new niche: A harsh, unkind drug would not hold appeal as a children's drug. Ritalin was therefore in perfect position to be in a "special" service as an MBD drug.

Educating Clinicians about MBD and Ritalin

Almost all the advertisements for Ritalin in the 1970s volumes of the *AJP* involve positioning Ritalin as *the* treatment for children with MBD. The advertising text declared that Ritalin had a "special role" in the treatment of MBD. The first advertisement in this series appeared in 1971. It was in part an educational advertisement, focusing on the need to legitimize the disorder and to inform clinicians about symptoms and evaluation of MBD. It was also educational in that this was the first advertisement to offer a representative child patient to clinicians. In the before-drug picture, this first Ritalin boy is reminiscent of the Mellaril boy we saw earlier: His behavior seems out of control. His face is angry—his drawn back mouth suggests that he might be shouting. One arm is raised as if to come banging down on the desk, the other hand claws at the closed notebook. The question above the picture on the first page: "medical myth?" contrasts the "reality" of the pathological behavior seen by the viewer; the juxtaposition of text and image challenges the viewer to deny the reality of this pathology—to keep "dismiss[ing]" the behavior as merely "youthful vitality." Once MBD becomes a "diagnosable disease entity" on the next page, order and calm are restored. The clinician is informed of the boundaries of diagnosis and of an effective treatment; the boy takes his pills and is once again manageable, under control.

This is the only advertisement in this series that shows the boy alone in all picture captions. He is in a school classroom, which provides an important interpretive context for his behavior. In this context, behavior that interferes with learning cannot be so easily dismissed. While there is no direct reference to boys specifically in this advertisement (all references are to "children"), the gendering of the problematic behavior is already present most clearly, of course, in the fact that the child depicted is a boy. But it is also present in more subtle descriptive ways; the reference to explanations that see this behavior as part of a child's "spunkiness," "youthful vitality," "aggression," and "mischievous behavior" suggest activities and an activity level that are cultural markers of boyhood. The behavior depicted in the before-treatment image provides an important contrast to these cultural understandings of "normal" boyhood, in that the boy is shown on his own, wrestling with invisible/internal demons that appear not to provide him any joy. The boy's lonely frustration interrupts cultural expectations of the sociability of normal boyhood vitality. And yet, the after-picture does not depict a "normal" boy either; rather, the emphasis is on the achievement of controlled behavior in the classroom. This controlled, contemplative boy, whose eyes are cast down, whose hands are neatly placed on the desk, whose hair is neat and tidy—he is not a "normal" boy either. The clinical expectation of treatment outcomes was not normalcy; rather, clinical expectation was built around the notion of contrast between behaviors pre- and post-treatment. Treatment did not normalize behaviors so much as it controlled them.

This emphasis on control allowed Ciba to shift the clinician's attention away from the problematics of evaluating normal/abnormal behavior in little boys, and to focus instead on the concrete possibilities inherent in Ritalin treatment. If earlier Ritalin adverts emphasized the gentle nature of the drug, and by association the prescribing clinician, this advertisement extended the holding ground to the vulnerable clinician himself. The opening text acknowledges him as a member of the community and invites him to join a circle of collective wisdom and expertise: "What medical practitioner has not, at one time or another, been called upon to examine an impulsive, excitable, hyperkinetic child?" The clinician's mistakes are simultaneously clarified and excused, in sotto third person passive voice: "In the absence of any detectable organic pathology, the conduct of such children was, until a few short years ago, usually dismissed." But now that the disorder is "readily diagnosed" and approved by an "expert panel," there need not be any more debate over the pathological etiology of ambiguous problem behaviors in young boys—and clinicians no longer need to worry about making more mistakes. There was now a means of controlling the behaviors and medical justification for doing so.

In subsequent 1970s adverts, there was more confidence in clinicians' knowledge of the MBD diagnosis and consequently less focus on the need to bring the clinician into a community of experts who diagnose MBD and prescribe Ritalin. The advert still spoke to the clinician-reader, of course, and the emphasis on controlling and managing behavior in a school context was sustained. What changed quite dramatically in the course of subsequent advertisements was the introduction of a third person into the dynamic between clinician-viewer and representative boy patient. At first, this person's status was somewhat vague; she could be a teacher; she could be a mother; she might be both. By the mid-1970s, she was definitively mother.

Insertion of the Female Figure

Advertisements from 1973 and 1974 introduced the importance of relationship into the depiction of problem behaviors in children and suggested that the relationship between mother/teacher and boy was an important factor in the representation of successful drug treatment. The sadness of "Wednesday's Child"—who is "full of woe"—is depicted with a picture of a crying boy who is closely held and comforted by an ambiguous female figure.

Here, Ritalin is a "helping hand"; the bold text closely aligns with the prominent female hand that embraces and shelters the boy, enclosing their relationship in a private space. But if the female hand is a visual barrier between the clinician-viewer and the boy, it also provides a set of metaphoric linkages between the "helping hand" of the drug, the helpful physician whose hand writes the prescription for the drug, and the woman/mother whose hand encloses the boy. All these hands collaborate in "managing Wednesday's child." And so, in fact, the impression of an enclosed relational space between woman/mother and boy is false. The advert encourages doctors to enter this space, to join their hands with that of the woman/mother, in order to embody and facilitate the drug's reach into the boy. Joined together, these actors promise to resolve not only the problem of individual behavior, but also the emotionally complex relational dynamics that accompany this "affliction."

A more straightforward representation of Ritalin's ability to manage woman/mother/teacher/boy relationships as well as individual behavior was demonstrated in an advertisement which showed a large picture of a vulnerable-seeming boy listening thoughtfully and carefully to an ambiguous woman who was assisting him with schoolwork. They are both engrossed in what they are doing; their focus isn't broken by the clinician-viewer. However on the ad's second page, the boy kisses the woman in a moment when they are not working (perhaps they have just come in from playing outside—the boy is wearing a hat). His kiss is in profile; his eyes are almost closed and his face is hidden behind her cheek. But her face and her smile are fully exposed, providing the only direct communication between the characters in the advertisement and the clinician-viewer. While the image on this second page is very small, its importance is clearly established in the happy kiss, which releases the focused tension evident on the previous page. That kiss, this relationship—these are key aspects of Ritalin's healing work.

In fact, this relationship is so important in this series of advertisements, that even when the advertisement is ostensibly about something else, a female/maternal presence is powerfully felt. Another 1974 advertisement depicted a doctor doing physical therapy with an MBD boy; the text encouraged the use of Ritalin alongside other treatments for MBD. The physical therapy session is taking place in a classroom, underlining once again the importance of the school context in interpreting and managing boys' behaviors. Apparently only the doctor and the patient are present in the room. However, at their backs, at the far end of the room, hangs a woman's winter coat. It is a curious place for a coat to be hanging, in a classroom, midway up against a window, facing out on a hanger as if to show its details, rather than simply hung on a coat hook behind a door. These curiosities, of course, indicate the importance of the coat's message: it asserts the presence of a woman in the room. Her identity is once again ambiguous.

This sustained ambiguity around the female figure in this series of advertisements may reflect the focus, in this early series of ads with children, on educating physicians about *management and control* of MBD children, rather than *normalization* of MBD children. The former was much more clearly and persuasively a clinical project, while the latter strayed into politically charged territory, where subjective factors endangered medical definition and classification. So the inclusion of women in these ads may at this stage have been carefully nuanced so as to manage the interpretive possibilities. The drug's claims and promises might have extended to the woman-boy relationship, but the contexts in which this relationship was depicted were not domestic contexts, rather they were school or learning-based contexts, in which behavior management was necessary and justified. To a certain extent such an approach to female representation in these adverts also allowed for a minimization of the importance or power of the woman-boy relationship; as long as the clinic defined the nature and context of this relationship, it held greater power over the boy than did the woman.

Focus on Mothers and Sons

It is only once the reluctance to fully depict mothers in Ritalin advertisements was overcome that the distinctions between a project of behavior management and a project of normalization became blurred. And yet in this blurring, there was also a much clearer picture of how the construction of behavior problems in boys was grounded in a set of interconnecting relationships among boy, mother, and clinician. In a 1975 advertisement that brilliantly foreshadowed the direct-to-consumer advertising campaigns that appeared two and a half decades later, these relationships were cast in a drama that acknowledged the interpretive importance of contexts beyond the school and clinic. Within the domestic and social contexts (including school) suggested in this advertisement, Ritalin managed the appearances that accompanied a particular lifestyle.

The before-image is a photograph of a family—or an attempted photograph of a family. There are two gleaming children, and one "difficult child"—the child with MBD. His behavior disrupts both the taking of the photograph and the portrait of an ideal family—bright, attractive, well behaved, in control, White, middle class. But it is not just the image of the boy that disrupts the idyll; it is also the image of the mother wrestling with the boy, an awkward, appealing look on her face. She is trying hard, doing her best to appear competent and attractive, but her position is literally skewed by this difficult son; in contrast to her good children's straight backs and perfectly positioned hands, she is "distraught," imbalanced by the aggressive embrace with this boy.

Thus, the before-image—both visually and in the text—makes a direct link between the behavior and appearance of children to the behavior and appearance of the mother and suggests that the clinician's tool, Ritalin, can help manage all these appearances. It is the first advertisement to specifically claim the domestic space and domestic relationships as a legitimate domain of treatment. To intervene in the domestic space, the expert and his tools necessarily confront maternal authority, and in the process of claiming the space must involve a reduction of mother's competence and power. In this advertisement, mother's reduction is affected both via the obvious inability of the mother to handle this boy on her own, and by the subtle insinuation of her role in the boy's problematic behavior. The text of the after-image tells the viewer that with Ritalin treatment, the boy is "a regular fourth-grader, *accepted at home*" (my emphasis). Embedded in the latter treatment result is a quiet reference to the well-established psychiatric literature on the pathologizing influence of rejecting mothering behavior on young boys.[39–41] In the mid-1970s, the problem of maternal acceptance of this child most clearly echoed John Bowlby's[42] widely read work on attachment disorders in children; and Bowlby's work was, of course, indebted to a psychoanalytic tradition that viewed the relationship between mothers and sons as inherently problematic and potentially damaging to the boy.

In this way, the pre- and post-treatment dichotomy depicted by this advertisement represented not only the successful management of the boy's behavior, but also the successful management of the mother's behavior—that is, returning the boy, the mother, and the family to the appearance of normalcy. It is notable that this advertisement does not promise actual normalcy; it promises only the appearance of normalcy—in photographs. The implication is that for the mother, who will be sending the school photographs to friends and family, the appearance of normalcy is highly valued and is possibly sufficient reason for Ritalin treatment. So, while this is the first advertisement to blur the distinction between Ritalin as a tool for managing behavior and Ritalin as a tool for normalizing behavior, there is still an element of self-consciousness, certainly reflexivity, in the use of a photographic metaphor to depict post-treatment normality. This reflexivity disappeared completely once Ciba began to advertise Ritalin directly to consumers.

We should not forget the perspective of the viewer-clinician in the analysis of this advertisement, although it is easier to forget him when viewing an ad that is so apparently devoid of clinical references. But there is a link, of course, between the perspective of the absent father (who is presumably taking the picture of his family) and that of the viewing clinician. Both men are invisibly positioned in an objective, evaluative posture as they gaze upon the mother and the situation. And there is a further link with the distressed mother, whose eyes make contact with the camera (her husband) and with the clinician-viewer who stands behind it/him.

The Politics of MBD and Drug Advertising

Why would Ciba believe that clinicians would respond well to an advertisement that so clearly focuses on improvement of lifestyle and appearance with Ritalin treatment? A possible answer may lie in the reference in the ad to the passage of time between 1971 and 1974. By the mid-1970s, there was an ever-increasing number of discussions and debates published in the popular media about psychotropic drug use for children's behavior problems. Congress was involved as well, through an investigation into the "Use of Behavior Modification Drugs on Grammar School Children" in Omaha, Nebraska.[43] The congressional hearings uncovered problematic and illegal prescribing practices which underlined concerns about the use of stimulant drugs as a form of institutional control of the individual.[44] Members of the subcommittee warned about the use of drugs to quiet the "bored but bright child." As this congressional committee went about its work, however, another branch of government, the NIH, was funding clinical research into the efficacy of stimulant drug use in children with MBD and other disorders (e.g. Eisenberg, 1964). Results of clinical research appeared alongside the public outcry during these years, with a stable and interconnected group of clinicians (several of them working from NIH grants) contributing to the work on stimulant drugs with children.[45] As a result, the early 1970s was a period of stabilization of knowledge claims in relation to both the MBD diagnosis and the use of stimulant drugs in children. In 1971, Paul Wender published a definitive text on MBD in children, with a foreword by Leon Eisenberg, who was at the time a prominent proponent of the diagnosis.[46] By 1974, an academic conference involving notable clinical and academic figures was dedicated to the subject of stimulant drug use in children. The conference resulted in a widely cited book.[47]

Thus, it would seem possible that by 1975, when this advertisement appeared, Ciba felt confident that clinicians would view MBD as a medical reality and would trust Ritalin as an appropriate treatment. Such confidence may have encouraged them to push the boundaries of treatment efficacy into a realm that was, of course, already implicated—Ciba did not create the conflation of the domestic and the clinical realms. The company's advertising did, however, draw expertly on this conflation, effectively grounding the evaluation and treatment of problem behaviors in boys in the interconnected relationships among mother, son, and clinician. This grounding in turn provided a marketing platform that allowed for the integration of medical tools and expertise into family life. Harmonious relationships could be represented as an important goal of drug therapy for the White middle-class nuclear family. Drugs could be sold on the basis of what they could "buy" in terms of acceptable social and domestic values. Chief among these values was mother, on whose emotional rationality and maternal competence familial and social harmony depended. Now mother could be managed, treated, and potentially transformed through her son's drug therapy. Thus, management of the mother, and by implication management of her relationship with the "problem" son, emerged as a key justification for medical intervention. The mother was indeed deeply implicated in Ritalin treatment.

Strangely enough, the Ritalin advertising trail, in the *AJP* and in other journals, went very quiet after this 1970s ad series. There was a brief and unsuccessful attempt to market a long-acting form of Ritalin in the 1980s, but for the most part, Ciba put its advertising money into antipsychotics. The only drug consistently advertised for ADHD children during this period was Cylert, made by Abbott.[48] It is difficult to know the reasoning behind the lack of Ritalin advertising in the 1980s, but one might speculate that the drug was well established for ADHD children by this time, and that Ciba needed to concentrate on launching and establishing other drugs. Moreover, ADHD discussion, debate, and possibly diagnosis also went very quiet in the 1980s, and there was a noticeable drop in the number of prescriptions for stimulant drugs during this period.[49] This lull helps to explain why the contemporary furor over ADHD and stimulant drug use in children seems so contemporary.

The Politics of Advertising Psychotropic Drugs

The 1980s were, however, important years for the pharmaceutical industry lobby. A series of congressional hearings, begun in 1957 by Senator Estes Kefauver, and continuing into the 1970s, had placed restrictions on pharmaceutical industry promotional activities. Psychiatric advertisements in particular had been targeted for providing misleading or insufficient information about both diagnosis and treatment.[50] During the Reagan years, however, ongoing developments in U.S. Food and Drug Administration (FDA) regulatory procedures eventually boosted the industry's ability to promote its products. Of particular relevance during this period was the gradual reinterpretation and revision of long-standing FDA policies prohibiting Direct-to-Consumer (DTC) activities for drugs. By the late 1980s, advertising of drugs in newspapers and magazines was finally allowed.[51] In 1997, advertising using broadcast media was allowed, and by 2001, the pharmaceutical industry was spending $2.7 billion on DTC advertising.[52] Despite these changes, the industry did not advertise stimulant drugs directly to consumers because of a 1971 United Nations Prohibition on Psychotropic Drug Advertising. While the United States did sign the Prohibition, no laws in line with the Prohibition were ever passed. Despite the fact that stimulant drugs are classified as controlled substances by the U.S. Drug Enforcement Agency (DEA), neither the FDA nor the DEA has the authority to control public advertising for Ritalin or other stimulant drugs for ADHD.[53] By 1999, the pharmaceutical industry had decided to act within its rights and launched the first DTC advertising campaigns for stimulant drug treatments for children. Given the historical roots of the preoccupation with mothers and sons both in child psychiatry and in public discourse, it is hardly surprising that this preoccupation continued to feature strongly when stimulant drug advertising came out of the clinical closet and into the mass marketplace.

Contemporary Direct-to-Consumer Advertising of Stimulant Drugs for ADHD

In stark contrast to the medical advertisements discussed thus far, contemporary Direct-to-Consumer ads for stimulant drugs make no visual attempt to educate the viewer about pre- and post-treatment behaviors. What contemporary ads depict is the drug's promise of life post-treatment. There are no symptoms, only solutions. Unsurprisingly, post-treatment images present a highly idealized portrait of a family life, in which everyone is attractive, middle-class, happy, and well behaved. It should be shocking to the viewer that people like this are also consumers of stimulant drugs. But of course, the message of the ad is that psychiatric diagnosis and stimulant drugs are in fact part of normal domestic life. But the life that is depicted is not normal. The journey of stimulant drug advertising into the public domain has taken the claims of the drug maker from control, to a blurring of control and normalization, to a clear vision of enhancement. Today the advertisements say to the female consumer: This is the kind of son, the kind of family, and the kind of relationships the drug will buy you. *You* will shine.

A majority of DTC advertising of stimulant drugs in the twenty-first century places the relationship between mother and son squarely in the visual and metaphoric center. The drug is positioned as a tool to manage this relationship; in the absence of symptoms, the mother-son relationship is the only suggested "problem" that needs fixing. Mothers who come across these ads, which are frequently placed in women's and parenting magazines, learn that successful mothers have successful sons, and that drug intervention is a sign of maternal love and care. They learn that stimulant drugs support a normal boyhood. Most important, they learn that harmonious nuclear family relationships—particularly the relationship between mother and son—are assured through drug intervention.

The ads provide a sense of this kind of DTC campaign. There are no symptoms, no pills, no problems pictured whatsoever. The boy never even looks up at the viewer; he is too engrossed in his good behavior or in his relationship with his mother. His focus is, of course, a desirable behavioral outcome of treatment. But the point of drug treatment, as depicted in this ad, is not simply good behavior; the real promise is an enhanced relationship between mother and son. Simple domestic scenes that were once problematic are now idyllic times of domestic harmony and partnership. Mornings, afternoons, and evenings with this boy are tranquil and controlled; the long-acting version of the drug means no more fights about dosing. The boy does what the mother wants him to do all day long; he is the perfect male companion (particularly in the absence of the father). The mother's dominance in this advertisement may be partially explained by the fact that such ads are targeting women readers. But the analytic interest is not so much that mothers appear in these ads; rather, it is the mode of mother representation that is remarkable. Her knowing, satisfied look out at the viewer (probably also a mother) is almost coy, alluring. It pulls the viewing mother into the frame with the promise that she too can share in the secret of this kind of relationship with a son. Why is the secret compelling to the viewing mother? Why does she want to buy this kind of relationship with her son? The drawing power of this ad is founded upon the reconstruction of a historical trope that links problem boys to problematic mothers.[54] The advertisers of Adderall appear to understand that this link causes mothers of boys extreme anxiety; they appear to understand that mothers are directly implicated in their sons' problem behaviors. And so they offer the modern mother a way out of this dilemma: a drug that will make the boy's problems invisible (as they are in this ad) so that mothers can look more like the stereotypical ideal. The trick is that the link between successful boys and successful mothers is just the other side of the link between problem boys and problematic mothers. On this side, the drug makers just make oppression look and feel better.

This link between successful boys and successful mothers is continued in an Adderall advertisement aimed directly at "parents of children with ADHD." As such, the drug is positioned explicitly as a problem solver for *parents* (read, mothers). Again, we might ask what the mother's problem is exactly, as depicted in an advertisement that reveals no child symptoms. What has "finally" come to pass post-treatment? What has the drug delivered for *the mother*? The answer is clear in the joyous hug, the connection between mother and son. This is what Adderall promises: Mothers will "finally" be able to hug their sons like "normal" mothers do, they will feel joy and pride instead of anger and resentment, and their sons will feel happy and loved. Note the test paper the boy is holding; he has been given a B+, a decent grade. The prominent red mark is meant to capture the viewer's attention. It confirms the goodness of the mother, telling us that she is not unreasonable in her desire for her son's success. She is not a pushy, competitive mother who puts her children on stimulant drugs so that their enhanced performance can feed her own ego. In contrast, she is overjoyed by a B+. This advertisement therefore not only flips the problem boys/problematic mothers equation onto the success side, but it also goes further to nuance the nature of the mother's engagement in a boy's success. In doing so, the drug makers may be responding to the cultural critique of the successful boys/successful mothers equation, which sees mothers as fostering their sons' success with stimulant drugs to meet their own needs. All this flip-flopping from one side of the cultural-historical trope to the other only underlines the reality that neither side of this trope is a safe or a liberating place for mothers to sit. But as long as the performance/pathology of mother is intimately linked to the performance/pathology of sons on a cultural level, this kind of advertising strategy will probably be effective.

This chapter raised a question about the phenomenon of ADHD and stimulant drug use: Why boys? One answer might be: Because mothers. This is put forward as a partial answer; it probably coexists with a number of others, including biogenetic and environmental factors that may be peculiar to boys. Most important, it is not meant to be a judgment of mothers. It is meant rather as a (re)assertion of the mother in Ritalin—and the psychoanalytic penumbra in biological child psychiatry—which cannot be overlooked if we are to understand how it is that so many mothers today pursue and desire expert intervention in their young boys' behaviors. Advertising plays a reflexive role in our society; it shapes us but it is also shaped by us. Its power lies in its ability to anticipate us—to discover our weaknesses and desires and use these to construct its promises, create more desire, and sell us products. The problematic terrain of young boys' normality and behavior embodies interconnected clinical and maternal anxieties; if we want clinicians and mothers to become more critical of industry promotional representations of stimulant drugs—so that decisions to treat children with these drugs can be made with greater autonomy and reflexivity—we need to continue to illuminate this anxious underbelly.

NOTES

1. David Healy's interviews with Judith Rappoport, Rachel Klein, and Leon Eisenberg make up a small, fascinating body of historical material on clinical experimentation with stimulant drugs. The interview with Judith Rappoport is in Healy, D. (2000). *The Psychopharmacologists vol 3*. London: Arnold, pp. 333–356; the interview with Rachel Klein is in the same volume, pp. 309–332. The interview with Leon Eisenberg (May 1998) is unpublished, but available from David Healy.

2. Rose, N. (2004). Becoming neurochemical selves. In Nico Stehr, ed. *Biotechnology, Commerce and Civil Society*, pp. 89–128. New Brunswick, NJ: Transaction Publishers, 2004.

3. International Narcotics Control Board (INCB). Use of methylphenidate for the treatment of attention deficit disorder. In: Report of the International Narcotics Control Board for 1995. UN Doc. No. E/INCB/1995/1:II.B.4. [http://www.incb.org/e/ar/1995/menu.htm]

4. Larry Diller, author of *Running on Ritalin*, was more successful in getting Novartis executives to talk to him. Novartis has not released much information about the development of the drug or about the clinical trials process.

5. Shrag, P. & Divoky, D. (1975). *The myth of the hyperactive child*. New York: Pantheon.

6. Moser, R. (1974). The continuing search: FDA drug information survey. *JAMA,* 229(10) 1336–1338.

7. McCue, J., Hansen, C., Gal, P. (1986). Physicians' opinions of the accuracy, accessibility and frequency of use of ten sources of new drug information. *Southern Medical Journal,* 79(4) 441–443.

8. Peay, M. & Peay, E. (1988). The role of commercial sources in the adoption of a new drug. *Social Science and Medicine,* 26(12) 1183–1189.

9. Orlowski, J. & Wateska, L. (1992). The effects of pharmaceutical firm enticements on physician prescribing patterns: There's no such thing as a free lunch. *Chest,* 102, 270–273.

10. Goenuel, F., Carter, F., et al. (2001). Promotion of prescription drugs and its impact on physicians' choice behavior. *Journal of Marketing,* 65(3) 79–90.

11. Kravitz, R., Epstein, R., et al. (2005). Influence of patients' requests for direct-to-consumer advertised anti-depressants. *British Medical Journal,* 293(16), 1995–2002.

12. Caudill, T. S., Johnson, M., et al. (1996). Physicians, pharmaceutical sales representatives, and the cost of prescribing. *Archives of Family Medicine,* 5(4) 201–206.

13. Cormack, M. & Howells, E. (1992). Factors linked to the prescribing of benzodiazepines by general practice principals and trainees. *Journal of Family Practice,* 9(4) 466–471.

14. Bower, A. & Burkett, G. (1987). Family physicians and generic drugs: a study of recognition, information sources, prescribing attitudes and practices. *Journal of Family Practice,* 24(6) 612–616.

15. The advertisements in this chapter were discovered through a systematic investigation of *The American Journal of Psychiatry,* from 1955 to 1988. Two additional medical journals, *Psychiatric Annals* and *The American Journal of Pediatrics,* were surveyed nonsystematically in order to discover any major differences in stimulant drug advertising campaigns in these journals. Direct-to-Consumer ads were gathered from American magazines in a nonsystematic manner between 2000 and 2004. Additional historical resources consulted as part of the investigation into the history of ADHD and stimulant drug use have been described in Singh, I. (2002). Bad boys, good mothers and the 'miracle' of Ritalin. *Science in Context,* 15 (4), 577–603.

16. It has not been possible to include a visual of all advertisements discussed in this chapter. Detailed descriptions are given in place of a picture where necessary. Readers who wish to view all the ads discussed in this chapter should please contact the author.

17. Charles Bradley was the Director of the Emma Pendleton Bradley Home in Rhode Island, the nation's first psychiatric home for children. Readers interested in a more extensive discussion of Bradley and the context of the Home are referred to Singh, 2002.

18. Bradley, C. (1937). The behavior of children receiving Benzedrine. *American J. of Psychiatry,* 94, 577–585.

19. Bradley, C. & Bowen, M. (1940). Amphetamine (Benzedrine) therapy of children's behavior disorders. *American J. of Orthopsychiatry,* 11, 92–103.

20. Bender, L. & Cottingham, F. (1942). The use of amphetamine sulfate (Benzedrine) in child psychiatry. *American J. of Psychiatry,* 99, 116–121.

21. Lauffer, M. & Denhoff, E. (1957). Hyperkinetic behavior syndrome in children. *Journal of Pediatrics,* 50, 463–475.

22. Singh, 2002.

23. Hall, A.L. (2005). Welcome to ordinary? Marketing better boys. *American J. of Bioethics,* 5(3), 59–60.

24. Metzl, J. (2003). *Prozac on the couch.* Durham, NC: Duke UP.

25. Smith, M. (1991). *A social history of the minor tranquilizers.* Binghamton, NY: Haworth Press.

26. Conrad, P. & Schneider, J. (1980/1992). *Deviance and medicalization.* Philadelphia: Temple University Press.

27. Kindlon, D. & Thompson, M. (1999). *Raising cain: Protecting the emotional lives of boys.* New York: Ballantine Books.

28. Pollack, W. (1998). *Real boys: Rescuing our sons from the myths of boyhood.* New York: Random House.

29. Metzl, 2003.

30. Singh, 2002.

31. Romano, S. (1998). A glaring oversight: The use of stimulant medications in child psychiatry without controlled testing. Unpublished paper.

32. Metzl, 2003.

33. Lauffer & Denhoff, 1957.

34. Connors, K. & Eisenberg, L. (1963). The effects of methylphenidate on symptomatology and learning in disabled children. *American J. of Psychiatry,* 120, 458–464.

35. Eisenberg, L. (1964). Role of drugs in treating disturbed children. *Children,* 2 (5), 167–173.

36. Eisenberg, 1964. Work cited in this 1964 paper was supported in part by a grant from the National Institutes of Mental Health, suggesting official acknowledgment of the importance of psychotropic drugs in child psychiatry.

37. "Those Mean Little Kids," *Time,* 1968; "Too Many Drugs," *Time,* 1961; "Pep Pills for Students," *Newsweek,* 1970.

38. Moynihan, R. & Cassels, A. (2005). *Selling sickness: How the world's biggest pharmaceutical companies are turning us all into patients.* New York: Norton Books.

39. Watson, J. (1928). *Psychological care of infant and child.* New York: Norton.

40. Strecker, E. (1946). *Their mothers' sons.* New York: Lippincott.

41. Fromm-Reichmann, F. (1948). Notes on the development of treatment of schizophrenics by psychoanalytic psychotherapy. *Psychiatry,* 11, 263–273.

42. Bowlby, J. (1970). *Child care and the growth of love.* (2nd ed.) Harmondsworth: Pelican.

43. Department of Health, Education and Welfare. (1971). "Report on the conference on the use of stimulant drugs in the treatment of behaviorally disturbed young children." Office of Child Development and the Office of the Assistant Secretary for Health and Scientific Affairs (January).

44. Shrag & Divoky, 1975.

45. Connors, C. K. (ed.) (1974). *Clinical use of stimulant drugs in children: Proceedings of a symposium held at Key Biscayne, Florida, 5–8 March 1972 (International Congress Series).* New York: Elsevier.

46. Wender, P. (1971). *Minimal brain dysfunction in children.* New York: Wiley-Interscience.

47. Connors, 1974.

48. Cylert was never widely used as a treatment for ADHD; it has been taken off the market.

49. Rose, 2004.

50. Smith, 1991.

51. Klawiter, M. (2006). The biopolitics of risk and the configuration of users: Clinical trials, pharmaceutical technologies, and the new consumption-junction. In S. Frickel & K. Moore, eds. *The New Political Sociology of Science: Institutions, Networks and Power.* Madison: University of Wisconsin Press, 432–460.

52. Angell, M. (2004). *The truth about drug companies: How they deceive us and what to do about it.* New York: Random House.

53. Novak, V. (2001). "New Ritalin ad blitz makes parents jumpy." *Time* (10 September), 62–63.

54. Singh, I. (2004). Doing their jobs: Mothering with Ritalin in a culture of mother-blame. *Social Science and Medicine,* 59, 1193–1205.

TRANQUILIZERS

Tranquilizers on Trial
Psychopharmacology in the Age of Anxiety

Andrea Tone

On May 22, 1978, some 17 million viewers of NBC's national evening news broadcast learned the medical misfortune of Cyndie Maginniss, a prescription drug addict. Like millions of other American women, the thirty-two-year-old wife and mother of three had struggled with the challenges of a busy life. When she discussed her difficulties with her gynecologist, he prescribed tranquilizers. When her problems got worse, he prescribed more. Maginniss soon discovered that she had become a prisoner of prescription pills, taking increasingly higher doses to keep calm. Breaking her tranquilizer habit proved difficult. "My body was completely out of whack," Maginniss told the NBC reporter. "Why did you wait so long before you got hooked?" the reporter asked. "I thought I was taking medicine," Maginniss replied.[1]

Maginiss's story represented one of hundreds of tranquilizer narratives recounted in newspapers, magazines, courtrooms, government investigations, and television studios across the United States in the 1970s and 1980s. Although the circumstances that turned anxious patients into drug addicts varied, these narratives offered a chilling account of medicines and patients veering out of control.

Only a few decades earlier, researchers, journalists, and physicians had hailed meprobamate, the first of the minor tranquilizers, as a triumph of American pharmaceutical science. Sold under the brand names Miltown and Equanil, meprobamate was flaunted as a safe and easy way to handle unproductive stress and anxiety. An article in the family magazine *Town Journal* typified the enthusiasm of the era. "Science has discovered an amazing new drug that effectively controls anxiety," it proclaimed. It "brings sleep to the sleepless, relaxation to the tense, tranquility to the nervous."[2] This kind of excitement had commercial and medical repercussions; by 1957, Americans had filled thirty-six million prescriptions for the drug. With sales topping the $200 million mark, meprobamate became the first psychotropic wonder drug in medical history.[3] By the 1970s, however, minor tranquilizers, which now included the best-selling benzodiazepine tranquilizers Librium and Valium, had been recast as dangerous drugs. They were recklessly prescribed, aggressively promoted, and carelessly consumed, less a sign of the triumph of pharmaceutical science than a commercial success achieved at patients' expense. In Washington, D.C., Senator Edward Kennedy opened a 1979 hearing on the "Use and Misuse of Benzodiazepines" by warning that tranquilizers had "produced a nightmare of dependence and addiction, both very difficult to treat and recover from."[4]

This chapter explores how and why this change occurred. It examines the history behind the development, medical diffusion, and political discrediting of tranquilizers. Its focus is twofold. First, it details the making of meprobamate and the attendant redefinition of anxiety as a medical problem that could benefit from pharmacotherapy. As we shall see, the drug's invention predated the creation of a clear-cut consensus that anxiety was a disorder requiring medical care. Nor did the drug's launch index a crafty commercial plan to ensnare a captive consumer market. Far from being a compound "rationally" designed for a waiting-and-wanting public, Miltown's development followed a haphazard path, one that elucidates the unexpected trajectories of medical inventions and the social construction of consumer markets. Second, this chapter looks at the reasons behind the seismic shift in popular and political perceptions of tranquilizers, examining how evolving ideas about culture, drugs, and social unrest reframed their place in modern medicine. In developing these themes, it seeks to contribute to the burgeoning scholarship on the history of psychiatry, psychopharmacology, and medicine by exploring the relationships between medicine and media, patient demand and therapeutic practice.

The Long March to Miltown

Beginning in the 1960s, critics of the tranquilizer bonanza accused pharmaceutical firms of rushing anti-anxiety agents to market, privileging profits over adequate testing, and seasoned reflection about the advisability of marketing daily tranquilizers at all. In this rendering of the past, business executives and doctors pushed tranquilizers into the mouths of an anxious nation. The evidence from the 1940s and early 1950s, however, tells a different story marked by a surprising degree of commercial reticence and medical doubt.

Meprobamate was the first of the minor tranquilizers and the brainchild of Dr. Frank Berger. Wallace Laboratories, the ethical wing of Carter Products of New Brunswick, New Jersey, recruited Berger in June 1949 to become its director of medical research. Berger was a Czechoslovakian-trained physician who had spent ten years in clinical and bacteriological research in his native country and England before arriving in the United States. While working on methods to extract and purify penicillin, he discovered the muscle relaxant properties of mephenesin (sold in the United Kingdom as Myanesin and in the United States as Tolserol).[5] A 1946 article in the *British Journal of Pharmacology* reported that its use in "mice, rats or guinea pigs caused tranquilization, muscle relaxation, and a sleep-like condition from which the animals could be roused."[6] The drawbacks of myanesin, mainly its short duration, precluded its widespread adoption in clinical medicine. But word of its benefits to patients suffering from preoperative anxiety and spastic and hyperkinetic disorders generated significant medical interest and raised the possibility of a tranquilizer for general use.[7]

Mephenesin can be considered one of the many "accidents of innovation" whose frequent occurrence in the history of medicine and pharmacy—the discovery of penicillin and Viagra are better-known examples—challenge the more comforting view of medical advance as the corollary to carefully coordinated scientific research and laboratory experimentation. Tranquilization was an unexpected side-effect of a compound developed to kill bacteria that destroyed penicillin. Berger himself has noted that "discoveries in medicine are often made in indirect, roundabout ways. Back in 1945 [when the muscle relaxant properties of mephenesin were discovered] I did not have any plans to discover either a muscle relaxant or a tranquilizer."[8] As popular and controversial as they would later become, tranquilizers began as a curious endnote to what at the time was identified as a far more pressing problem, which was to protect the antibiotic supply.

Nor did the availability of mephenesin generate urgent commercial interest in the development of a tranquilizer for broader use. There was no "race" for Miltown that paralleled the rush to develop other drugs such as the benzodiazepine tranquilizers. In fact, one of the ironies of this history is the realization that Carter Products was not sure that a tranquilizer for outpatient anxiety was even worth bringing to the market.

Today the drug industry is the most profitable sector of the American economy, but in the 1940s and 1950s, its fate was unknown. Many companies were still small outfits, struggling to get their financial bearings. Carter was among them. Founded in 1880 by general practitioner Dr. Samuel Carter, it was in 1949 a small firm with total pharmaceutical sales of $550,000.[9] Before Miltown, its best-known products were the Carter's Little Liver Pills brand of laxative, Arrid deodorant, and Nair depilatory. The company was still trying to break into the prescription drug market. This was no easy task given that Wallace Laboratories consisted of only a few rooms in the company's New Brunswick factory and a couple of chemists whose chief job was to perform quality checks on Little Liver Pills.[10]

Because Carter executives were unsure if Miltown fit the bill, the history of the drug's march to market is punctuated with curious stops and starts. On the one hand, executives had recruited Berger expressly to create a minor tranquilizer, hoping he would find a longer-acting and hence more profitable alternative to mephenesin. After testing some 500 new compounds, Berger and the company's chief chemist Bernard Ludwig met this goal, synthesizing meprobamate in May 1950. Yet instead of immediately acting on their discovery and pushing the drug to market, Carter executives took their time making meprobamate available. When Berger produced data demonstrating meprobamate's efficacy in minimizing states of anxiety and tension, Henry H. Hoyt, president of Carter Products, refused to fund large-scale clinical trials, the next step in the drug's launch.[11]

Hoyt's reaction may seem surprising given the subsequent profitability of minor tranquilizers. But in the context of the pharmaceutical economy of 1950 it made sense. There was no preexisting market for prescription-only tranquilizers, and no one could predict how they would perform. The use of mephenesin had been largely hospital-based, whereas meprobamate was intended to be a drug for use in noninstitutional patients too. In addition, in 1951, Congress had passed the Durham-Humphrey Amendment to the 1938 Food and Drug Act, which meant that meprobamate could be available only by prescription.[12] This was new territory for drug companies, and Hoyt worried that physicians would be reluctant to endorse a drug for run-of-the-mill tension. Although some critics would regard physicians as accomplices in a pharmaceutical cabal—pushers of companies' crude "a pill-for-every-problem-syndrome"—Hoyt and his cronies considered them independent agents whose loyalty would have to be wooed and won.[13]

Adding to company concerns was the fact that the commercial possibilities for a psychotropic agent seemed bleak. Psychiatry departments throughout North America were still enamored with Freudian theories of neurosis, which taught doctors to regard anxiety as the product of unresolved conflict stemming from the unconscious mind. So ingrained was the Freudian tradition that in 1956, notes historian Nathan Hale, John Crosby could write that psychoanalysis was about as "controversial as the American flag."[14] Biological psychiatry had not yet come into vogue. While anxious persons consumed vast quantities of bromides and barbiturates, the prevalent medical model favored psychoanalysis, where personal history rather than brain chemistry was the therapeutic focus.

Hoyt's reservations were compounded by the company's internal data. Wallace had commissioned the Harris Poll of Princeton, New Jersey to conduct a survey of 200 doctors to gauge interest in a prescription anxiolytic. A majority of responding physicians expressed little interest in such a drug, and Hoyt wondered whether a medical market for anxiety was even possible. Unwilling to risk the possibility that it was not, he refused to give Berger the $500,000 needed to complete large-scale clinical evaluations to satisfy FDA requirements. What would later be the best-selling psychotropic drug in American history was, for the time being, shelved.[15]

Berger persevered, searching for ways to change medical minds. In a move befitting the cultural temperament of 1950s America, he and Carter colleague Thomas Lynes made a movie showing the effects of meprobamate on Rhesus monkeys in Berger's lab. The Miltown film featured monkeys in three distinct chemical states: naturally vicious, unconscious on barbiturates, and calm but awake on Miltown. Premiering at the 1955 meeting of the Federation of American Societies for Experimental Biology in San Francisco, the film caught the attention of executives from the rival drug firm Wyeth. What worked for monkeys presumably worked for men. Wyeth offered Carter a hefty sum to license meprobamate under the trade name Equanil. With Wyeth offsetting the cost and risk of bringing the drug to market, Carter followed suit. Carter's Miltown became available in May 1955, a full five years after Berger and Ludwig had concocted their compound. The march to Miltown had been long indeed. And, on the eve of meprobamate's release, no one was entirely certain how the drug would be perceived.[16]

Miltown Mania

Carter's Miltown and Wyeth's Equanil quickly became a commercial sensation, the first psychotropic wonder drug in history. Only fourteen months after it was made available, meprobamate had already become the country's largest-selling prescription drug.[17] By 1957, meprobamate had become the fastest-growing drug in history, selling at a rate of one prescription per second.[18] How do we explain Miltown mania? What made mild tranquilizers so popular so quickly?

The answer lies in a convergence of events and ideas, contingent to mid-1950s America. One was the heady optimism that informed attitudes toward new drugs. The decade as a whole was characterized by widespread faith in the possibilities of pharmaceutical science. Companies had recently made available a host of new prescription drugs, including antibiotics and synthetic hormones. Children who survived what a few years earlier might have been a deadly bout with bacterial pneumonia, adults with rheumatoid arthritis who were liberated from wheelchairs because of cortisone: these well-publicized triumphs provided tangible evidence of the wonders wrought by pharmaceutical medicine. No illness seemed beyond science's reach.

Nor was there, as yet, reason to doubt the safety and efficacy of prescription medications. The thalidomide tragedy and DES disaster had yet to occur. In a society where "ethical" drugs enjoyed cultural currency and reports of drug-induced injuries were infrequent, the debut of an antianxiety pill induced much excitement.

This spirit of optimism framed popular responses to Miltown's release. *Cosmopolitan* touted the new pills as "just the thing" for "tension-ridden, nervous patients: perfectly normal people who need temporary help."[19] The Hollywood tabloid *Uncensored!* reassured patients that they could take Miltown and Equanil with confidence because "they are not habit-forming and even a severe overdose can't kill you."[20] Word from prominent doctors was equally sanguine. The esteemed psychiatrist and researcher Dr. Nathan Kline, director of New York's Rockland State Hospital, told readers of *Business Week* that in its potential benefit to Americans,

the advent of minor tranquilizers was "equal in importance to the introduction of atomic energy, if not more so." Meprobamate restored "full
efficiency to business executives" and put artists and writers "suffering
from long periods of nonproductivity because of 'mental blocks'" back on
track.[21]

Physicians immersed in this culture of pharmaceutical excitement
could not help but be affected by it. Their willingness to prescribe
meprobamate and other minor tranquilizers was further cultivated by
four parallel developments: the successful introduction of major tranquilizers, the publication of scientific articles on meprobamate, pharmaceutical advertising campaigns, and patient demand. In 1954, a year before
meprobamate's debut, McGill university psychiatrist Heinz Lehmann
transfixed the psychiatric world with reports of the first clinical trial of the
drug chlorpromazine, which Lehmann had given to seventy patients with
schizophrenia, severe depression, and organic dementia. The drug produced remarkable results. In a matter of weeks, patients who had previously entertained no hope of recovery or discharge were symptom-free.
Word of Lehmann's study made its way to the United States, where chlorpromazine (sold under the brand name Thorazine) was quickly adopted
in mental health institutions such as Pilgrim State Hospital in New York,
the nation's largest. Tales of miraculous remissions and unexpected dis-
charges spread like wildfire. No one doubted the drug's significance; as
Lehmann himself put it, the coming of chlorpromazine represented nothing less than "the most dramatic breakthrough in psychopharmacotherapy
since the advent of anesthesia more than a century before."[22] The widespread adoption and demonstrated value of antipsychotic agents such as
chlorpromazine and the drug reserpine, soon known as the "major tranquilizers," lent credence to the claim that medications might benefit other
psychiatric disorders too.

Even as major tranquilizers were stirring up interest in psychotropic
medicines, in general, the first studies of meprobamate were published in
medical journals. In 1955, Dr. Edward Schlesinger of Columbia University
and Dr. Lowell S. Selling of Orlando, Florida, published the results of the
first two large-scale clinical trials of meprobamate. Schlesinger and Selling's articles, which appeared in the prestigious and widely read *Journal of
the American Medical Association*, reported that 78 to 80 percent of
patients suffering from tension or anxiety states improved or recovered
after using meprobamate. According to Borrus, the drug resulted in a
"lessening of tension, more complete relaxation, more prolonged restful

sleep, and ability to feel at ease in groups and when speaking before groups."[23] The well-placed articles, which would be followed by other equally positive reviews, helped build medical confidence in the drug.

To disarm critics worried that psychopharmacology would displace psychoanalysis, drug manufacturers orchestrated an artful promotion that highlighted meprobamate's applications as an adjunct to psychotherapy. In 1950s American outpatient psychiatry, neurosis was the favored diagnosis; as David Healy has shown, today's more popular diagnosis of depression came into vogue much later.[24] Psychiatrists treated neurosis with long-term analysis grounded in introspective reflection, helping patients through increasingly uninhibited conversation to peel back the layers of repression and inhibition that masked the "true" and early etiology of adult disorders. On the surface, this approach left little room for tranquilizers; indeed, drugging the brain seemed fundamentally at odds with the personal and subjective orientation of talk therapy. To collapse this dichotomy, drug companies positioned minor tranquilizers as just another tool in the analyst's arsenal, a chemical aid that would enable psychiatrists to relax their patients and identify the crux of their problems sooner. An advertisement for Miltown in the *American Journal of Psychiatry* showed a psychiatrist, pen in hand, consulting a young man whose face was seized with worry. Bold letters delivered the advertisement's main message: that Miltown "improves rapport when anxiety blocks therapeutic progress in private practice." Miltown merely helped the psychoanalyst do his (reflecting the profession's gender demographics, the doctor in Miltown ads was always male) job better. It "reduces both overt and covert anxiety levels and helps the patient overcome neurotic inhibitions. It improves patient cooperation and facilitates productive sessions."[25] The company's hefty *Physician's Reference Manual* underscored the point, reinforcing both the importance of anxiety as a medical disorder and the centrality of the psychiatrist's therapeutic role.[26]

As drug companies actively courted psychiatrists' support, patients were demanding meprobamate from their doctors. If one doctor was unwilling to give them a prescription, patients could always find another who would. To view the diffusion of Miltown as a top-down process turns patients into pawns and occludes instances of patient influence and negotiation that, as historian Roy Porter and others have argued, have always given medicine its dynamic character and shape.[27] In the case of minor tranquilizers, it would be misleading to ignore patients' involvement for, from the very beginning, they were the drug's greatest enthusiasts. Although the idea of a tranquilizer targeted for anxiety was new, meprobamate did not arrive in a pharmaceutical vacuum: thousands of Americans were already taking habit-forming and dangerous barbiturates when meprobamate, purportedly unaddictive and safe—"suicidal attempts with Miltown have been unsuccessful despite the ingestion of large amounts of the drug," claimed Carter—came into view.[28] Patients sufficiently worried by the well-hyped hazards of barbiturates, a class of drugs a 1951 cover story in the *New York Times* called more addictive than cocaine, changed their prescriptions.[29] Meprobamate also created a new drug-taking population among people who years earlier might have eschewed pharmaceutical remedies altogether. The inducements to do so were myriad, from the country's giddy pharmaceutical enthusiasm to doctors' frequent reassurances that tranquilizers were beneficial, effective, and safe. As well, Americans in the mid-1950s had no reason to doubt that these were, indeed, anxious times. Suburban bomb shelters, duck-and-cover drills, and dog tags for children were not only daily reminders of what might happen in a Cold War world, but also an invitation to be more than passive bystanders to its vicissitudes. If anxiety was disarming in its ubiquity and impossible to eradicate in its entirety, it was also something one could manage and contain. In the absence of an organized anti-drug voice, the more reasonable question in many Americans' minds in the 1950s might very well have been "why not take a tranquilizer?"

In the largest of cities and smallest of towns, Americans who had read about the new pills in newspapers or magazines or learned about them from family and friends visited doctors to get a supply. The "worried well" consulted doctors they were most likely to see: not psychiatrists but family practitioners, internists, pediatricians, and obstetrician-gynecologists. From the very beginning, then, this critical chapter in the history of psychiatry and psychopharmacology, the explosive rise of psychotropic medicine, was carried out by nonpsychiatrists. The success of new drugs helped promote the rise of biological psychiatry: the belief that imbalances in the brain cause mood and affective disorders. Yet even as it shored up the credentials of psychiatry, the chemical revolution undermined the importance of formal psychiatric training by transferring the practice of everyday psychiatry from the specialist's office to the generalist's prescription pad. By 1975, just 25 percent of prescriptions for minor tranquilizers were written by psychiatrists.[30] In the age of Miltown, the medical management of anxiety became a mainstream affair.

Therapeutic "success stories" nurtured the middle-class movement for Miltown. In Hollywood, "Miltown" Berle, a daily devotee, assured millions of television viewers that they were addicted only if they were taking more than their doctors.[31] One magazine told the tale of a thirty-six-year-old salesman with acute job-related stress. The man's "tremors, insomnia, restlessness and anxiety forced him to quit." The man consulted a psychiatrist who observed his "constant agitation." The patient "could not sit still for more than a few minutes without having to get up and pace." After other drugs failed to ease his worries, the psychiatrist prescribed meprobamate. The results were miraculous. The salesman "telephoned to cancel further appointments, saying that for the first time in his life, he was free from a constant feeling of shaking and tension and that he was going back to work."[32] Here was a true American success story: with a little pharmaceutical help, the neurotic unemployed had once again become a productive breadwinner. In another case, the president of a Madison Avenue advertising agency produced a large bottle of Equanil and invited colleagues to "join in." They did. According to one observer, "the meeting went off with less argument than any they had held in years."[33]

Not only businessmen but also housewives, athletes, celebrities, and even doctors praised their "tranks." While minor tranquilizers would later be regarded as mother's little helpers, they were in the 1950s used by a wide swath of the population. Indeed, the drug's fanatical following among businessmen in this decade earned it the nickname "Executive Excedrin." Pharmaceutical companies encouraged doctors to prescribe meprobamate for men, for they had no financial incentive to confine tranquilizers to half the population. The surest path to profits was to position tranquilizers as a drug suitable for all anxious Americans. In pharmaceutical advertising in the 1950s, the gender of the "anxious" patient was just as likely to be male as female—a point overlooked by scholars who view the history of psychiatry and advertising chiefly as a story about women.[34] Moreover, from the patient's perspective, there was as yet no reason for men to feel awkward or embarrassed about taking tranquilizers. As the "Executive Excedrin" label suggests, tranquilizers were very much a man's drug. The stigma of mother's little helper, which linked the feminization of prescription drug-taking to societal decay, had not yet taken hold. In such a gender-neutral milieu, men remained avid users and outspoken proponents, effectively stoking patient demand.

Consumer requests for meprobamate in the 1950s were so pronounced that demand quickly exceeded supply. Within months of the drug's appearance on the market, script-toting patients hungry for made-to-order tranquility emptied drugstores of their stock, forcing startled pharmacists to post signs reading "out of Miltown" and "more Miltown tomorrow.[35] In one California drug store, pharmacists dealt with the high volume of scripts by rationing their supply, filling only half of a prescription at a time.[36]

Manufacturers were caught off guard by this grassroots frenzy. "Our inventory," admitted a Carter official in May 1956, "is zero. We're working overtime, and we're shipping tablets just as fast as we can package them, but we can barely fill our orders."[37] Frustrated patients sought pharmaceutical relief from the emerging bootleg market. In New York City in 1956, a tablet of meprobamate cost 9 cents with a prescription and 42 cents or more without.[38]

The popularity of minor tranquilizers fomented a revolution in how people viewed and used prescription medicines. It encouraged Americans to decide that it was "okay" to see doctors for drugs not to cure a disease but to make them feel better about living in the world. It made possible the revolution in lifestyle drugs that followed. Pharmaceutical executives awestruck by meprobamate's success gambled that if people would line up to buy drugs for anxiety, they might buy pills for other problems too: depression, difficulties concentrating, a weak libido. Prozac. Ritalin. Viagra. Each owes something to the Miltown moment, when anxious people reached for their pocketbooks and cashed in on the elusive promise of better living through a pill.

What tranquilizers did not do—at least at first—was threaten the social order. Their enthusiastic reception must be understood by looking at how critics calculated the drugs' risk not only to individuals but also to society as a whole. Tranquilizers seemed to smooth over cultural fissures that, left unattended, had the potential to become something more serious. They coaxed unemployed men back into the workplace. They made breadwinners, artists, and writers more productive. They softened the complaints of exhausted mothers. They made juvenile delinquents less edgy, soothed crying babies, and quieted agitated prisoners. As Jonathan Metzl has argued, minor tranquilizers were at once medically innovative and culturally conservative. Containing festering anxiety, they promised to keep the social fabric whole.[39] There was no "tranquilizer problem" in the public's eye until this important link between tranquilizers and social stability, forged in the 1950s, came unglued in the 1970s.

Benzodiazepine Blockbusters

Until that time, use of minor tranquilizers escalated. The popularity of meprobamate paved the way for Librium and Valium, two minor tranquilizers that belonged to a new chemical class of drugs called the benzodiazepines. These were drugs that worked chiefly as anti-anxiety agents and muscle-relaxants. Unlike mephenesin, discovered by accident, and meprobamate, marketed only after a crisis of commercial confidence, Librium and Valium were the results of a global race to identify a drug that would outsell Miltown. Texas psychiatrist Irvin Cohen, who ran clinical trails on Librium, later observed that the benzodiazepines were a "model of how a therapeutic agent is conceived and brought forth by an enterprising pharmaceutical manufacturer who simply seeks to find a drug superior to others already in the marketplace."[40]

The first benzodiazepine was Librium (chlordiazepoxide), synthesized by organic chemist Leo Sternbach in 1955 at the Nutley, New Jersey-branch of Swiss-based Hoffman-La Roche. Less sedating and more potent than meprobamate, Librium was tested on mice and then lynxes in the San Diego Zoo, where it was shown to "transform fierce wild cats into tame kittens." In clinical trials, Librium soothed but did not sedate outpatient "neurotics" and male prisoners.[41]

In March 1960, Librium was marketed amid a wave of positive fanfare. Roche promoted its new drug aggressively, spending over two million dollars in Librium's inaugural year to hype it to physicians in medical journals, direct mail, and pharmaceutical detailing. The advertising campaign trumpeted two main points. The first was Librium's status as a new chemical agent and not "a manipulated molecule." Many of the roughly 400 new drugs introduced each year at this time were variations of older drugs. Librium was, indeed, chemically new. Roche made much of this distinction. Its tranquilizer was a "unique product, different from all others that came before."[42]

The second strand of Roche's campaign was Librium's suitability for patients with moderate anxiety states. The tranquilizer market had hitherto been divided into two therapeutic spheres: meprobamate for mild anxiety and antipsychotics such as chlorpromazine for severe emotional disorders. After two years of clinical trials, Roche determined that Librium would help patients whose anxiety ranged "from the mild to the moderately severe," that is, patients whose symptoms straddled either sector. This finding encouraged Roche to pitch the drug to a larger portion of the tranquilizer market.[43]

Doctors and patients quickly accepted Librium's superiority over other tranquilizers. Three months after it was approved, Librium had become the most prescribed tranquilizer in the nation, giving newcomer Hoffman-La Roche a 16 percent share of the coveted U.S. prescription drug market.

Reports of Librium-related problems soon surfaced. The most important of these was published by Leo Hollister in *Psychopharmacologia* in 1961.[44] Roche had invited Hollister, an employee of the Veterans' Administration in Palo Alto, California and a member of Stanford's medical faculty, to a 1960 meeting convened to discuss Librium. Attending as an impartial observer, Hollister became concerned after listening to the parade of glowing endorsements. "If this drug is as good as these people say," he remembered, "it's going to be abused." In particular, Hollister worried about the drug's ability to create problems of physiological dependence or addiction.[45]

Hollister decided to study the issue by administering large doses of Librium to schizophrenic patients. His test was designed to assess both the drug's usefulness for patients with serious psychiatric disorders and its potential to cause withdrawal symptoms when it was abruptly discontinued. For several months, Hollister gave thirty-six hospitalized patients "monumental doses" of Librium. Eleven patients were put in the withdrawal study. Switched to placebo, ten of the eleven suffered withdrawal reactions: insomnia and agitation, lack of appetite, and nausea. Two patients had seizures. Hollister's study demonstrated that it was possible to become physically dependent on chlordiazepoxide in high doses. This important finding meant that Librium had the same dependency potential as barbiturates and meprobamate. The main difference was that "symptoms from withdrawal of chlordiazepoxide were slower to develop and less acute than those following withdrawal of barbiturates or meprobamate." Nevertheless, they were definite and consistent with a withdrawal reaction.[46]

Hollister's findings effectively countered manufacturers' claims that even the new-and-improved tranquilizers were not habit forming. Among psychiatrists, Hollister's study was broadly discussed and debated. "It was a study everyone talked about at the time," one psychiatrist recalls.[47] Subsequent pharmacological research investigated a range of dependence-related questions. How high a dose was necessary before patients experienced withdrawal? Did withdrawal occur only after long-term use? Were all patients equally susceptible or were some predisposed to dependence?

Concern about drug dependence in the psychiatric community failed to find much of an audience in other medical circles, however. The prescription addiction issue in general, and tranquilizer dependency in particular, did not make its way into mainstream medical journals, the very literature the majority of prescribing physicians were most likely to encounter. Even among physicians aware of the problem, the benefits of a drug that reduced social conflict often outweighed its risk of individual dependence. There was much to be said for keeping patients calm and content. And enthusiasm for psychotropics was still rampant; by the end of the 1950s, psychopharmaceutical research had yielded two Nobel prizes. In this milieu, patients kept demanding it, doctors kept prescribing it, and the media stayed mum about a problem as grim-sounding as addiction.

For Roche, these were happy times. Three years after Librium hit the shelves, the company outdid itself with Valium (generic diazepam). More potent than its predecessor, Valium was ten times more effective as a muscle relaxant and anticonvulsant. Roche poured millions into promoting Valium as the most versatile of the minor tranquilizers. One conservative estimate placed the cost of Roche's Librium and Valium campaign at $400 million. More revealing was how the sales income was used. Only 2 percent financed manufacture and distribution. The rest—fully 98 percent—went to profit and promotion.[48] Valium was known as "Valium the versatile." It could be prescribed for almost anything. Its chief indication was anxiety, but it also helped patients with muscle strain, alcohol withdrawal, gastrointestinal problems, cerebral palsy, multiple sclerosis, Parkinson's disease, and stroke. Between them, Librium and Valium cornered the tranquilizer market, accounting for $200 million of Roche's $280 million in sales by 1971. *Fortune* called the benzodiazepine blockbusters "the greatest commercial success in the history of prescription drugs."[49] Valium was the most widely prescribed drug of any kind in the Western world between 1968 and 1981. At the drug's sales peak in 1978, Hoffman-La Roche sold nearly 2.3 billion tablets, enough to medicate one-half the global population.[50]

Marketed as suicide-proof, Valium was considered safe, even when used in excess. Valium rapidly became a staple in medicine cabinets, as common as toothpaste, brushes, and razors. In what the poet W. H. Auden called the "age of anxiety," Westerners had found their favorite chill pill.[51]

Addiction by Prescription

The mood changed in the early 1970s. Once embraced as drugs no part of the civilized world should be without, tranquilizers began to be criticized for their adverse effects, particularly their addictive character.

That this change in how tranquilizers were perceived occurred against a backdrop of continuity reminds us of the importance of social and political variables in shaping a drug's fate. By 1970, the penchant for prescription pill popping was a well-established tradition and taking mood-altering drugs a decades-old habit. Hollister's study of benzodiazepine dependence had been in print for almost a decade. Other dependence studies, including another by Hollister on diazepam, had been published. What had changed, in short, was not the fact that Americans took tranquilizers—a whopping 800 tons in 1977—or that, like Cyndie Maginnis, many had a tough time when they stopped.[52] What was new in the 1970s was that Americans were for the first time anxious about their prescription drug behavior.

In the late 1960s, the United States was engulfed by worries that the cultural and therapeutic popularity of anxiolytics such as Valium made worse. The thalidomide tragedy and nagging doubts about the safety of oral contraceptives had burst the bubble of confidence in pharmaceutical panaceas. Patients across the country contemplated the extent to which drugs could hurt as well as heal. Thousands got caught up in the burgeoning women's health and consumer movements, which shared common ground in their suspicion of pharmaceutical and medical interests and their demands that patients be better apprised of the risks of all medical technologies, including drugs.

Then there were the illicit drugs, such as marijuana and LSD, that were being widely used by a rebellious youth. What historian Theodore Rosak called the "counterculture" was kindled by the coming of age of the baby boomers. By 1970, the cohort of Americans between ages 15 and 24 represented almost eighteen percent of the population, a postwar high. Many were well-educated; fully half of the 18 year olds went to college. Many openly defied materialism, consumerism, and the political leadership that had pushed the country into Vietnam. There were often no clear boundaries separating how people protested. Some channeled their resentment into student sit-ins, the anti-war movement, civil rights marches, and feminist rallies. Others registered their skepticism by embracing free love, psychedelic art, and the sensuality and driving political messages of the music of the age: rock. Very often, however, protests involved experimenting with illicit drugs—illicit not only because, like marijuana, they were illegal, but also because the social contexts in which they were used threatened political stability. There were hippies who dropped acid while living in communes, denouncing the tyranny of property ownership. There were angry Vietnam veterans who returned to the United States with an unwanted heroin addiction and barbed words about American imperialism in Southeast Asia. In January 1967, thousands of people gathered in San Francisco's Golden Gate Park for the first "Human Be-In." They swayed to the music of the Grateful Dead and Jefferson Airplane, listened to Beat poet Allen Ginsberg's chanting of Buddhist mantras, and heard Timothy Leary, a psychologist who had abandoned a promising career at Harvard ("LSD is more important than Harvard") to promote the psychedelic cause, tell the crowd to "drop out, turn on, tune in."[53] LSD had legitimate therapeutic applications. It had been used to help alcoholics and schizophrenics. But in the 1960s, as historian Erika Dyck has shown, it was discredited as "an agent of the counter-culture" and its manufacturer, Sandoz, halted production. The moral panic over drugs effectively doomed a compound with wide-reaching therapeutic potential.[54]

But it was not just LSD and "psychedelic psychiatry" that were scrutinized; mind-altering drugs of all kinds were newly regarded with suspicion. Tranquilizers previously lionized for their ability to patch social fissures were disparaged as disruptive and dangerous. Like Leary's disciples, tranquilizer takers were "tuning out" the realities of the world. That users got their drugs from doctors rather than on the streets made them no less dangerous. Indeed, it made the "drug problem" that more far-reaching and ominous, for it meant that mind-altering agents had penetrated the "safe" inner sanctum of middle-class suburbia.

By the early 1970s, the media had seized on this image of chemical contamination to promote the idea of widespread but secret middle-class addiction. "Condemnation of drug abuse has been primarily directed against hippies and narcotics addicts," Leonard S. Brahen, the director of medical research and education in the Nassau County department of drug and alcohol addiction, told participants in the 1972 meeting of the New York State Medical Society. "We now recognize the abuse of mood changing drugs is more extensive, involving people in all socioeconomic classes. Most middle-class housewives use such agents legally and consciously. . . . There is [even] evidence that a smaller number of suburban housewives have experimented with marijuana."[55]

In this cacophony of concern, nothing was more disturbing than the realization that stay-at-home moms, sentimentalized symbols of wholesome family values, were drug users. "The typical woman who uses drugs to cope with life is not a fast-living rock star, nor a Times Square prostitute, nor a devotee of the drop-out-and-turn philosophy," reported the *Ladies' Home Journal* in a 1971 exposé, "Women and Drugs: A Startling Journal Study." "She is an adolescent, confused by the stresses of impending adulthood. She is a newlywed, by turns anxious and depressed by strains of adjustment to a new relationship and new responsibilities. She is a once-busy housewife, her youngsters grown, who finds her days increasingly empty and her thoughts obsessed with the inexorable passing of the years." She was, in short, "an average, middle-class American—one of the folks next door. She could even be you."[56]

Women as a group consumed twice as many minor tranquilizers as men, but studies showed that it was non–wage-earning women aged 35 and over who consumed the most. A divorcee from Topeka took them for insomnia. "It does the trick," she told a reporter. A Des Moines housewife who got nervous at parties popped a Valium before heading out to put herself at ease. A mother of five found tending to her kids imposed "a lot of pressure." She took four pills a day and was grateful for the relief. "I never feel jittery."[57]

Some researchers made these findings part of a broader feminist critique of society's mistreatment of women. The real problem, they averred, was the circumstances that led women to seek escape. Why, indeed, should a woman feel calm minding five children alone? How fulfilled could a woman be cleaning dirty floors and toddler spit-up? Why did society expect women to look and act a certain way? Tranquilizer use was a logical corollary to an unhealthy ordering of gender roles. Canadian sociologist Ruth Cooperstock found in her research for the Addiction Research Foundation that wage-earning women were significantly less likely to take tranquilizers than were women who stayed at home. Cooperstock suggested that employment had a positive effect on women's well-being. It made them less anxious and "less apt to commit suicide." Sociologist Pauline Bart of Chicago's Abraham Lincoln School of Medicine reported that women often got locked into roles that made them unhappy. Isolated at home, they felt powerless to change their situations. "Instead of getting rid of the real constraints in their lives," Bart said, women take drugs "so they'll be better able to bear the pain."[58]

In a different time and place, these analyses might have fomented a political discussion of the mistreatment of women and a blueprint to help housewives experience their full potential as humans. But America's anti-drug fervor favored simple rationalizations over complicated explanations that shared the blame. Visionary thinking on this issue was left to the feminist movement. The mainstream media instead projected a straightforward and ultimately more reassuring message that discounted women's grievances and instead blamed doctors and companies for "pushing" dangerous drugs.

"Drug Abuse—Just What the Doctor Ordered" and similar headlines depicted physicians as ignorant puppets of the avaricious drug establishment. Doctors were, in the words of one journalist, prescribing tranquilizers "promiscuously."[59] The publication in 1979 of CBS producer Barbara Gordon's best-selling *I'm Dancing as Fast as I Can* encouraged this view. Gordon recounted how her psychiatrist hooked her on Valium only to recommend later that she withdraw "cold turkey." Gordon followed his advice and suffered from severe withdrawal symptoms. She also experienced psychiatric distress incapacitating enough to land her in a New York psychiatric asylum. Gordon blamed her doctor and Hoffman-La Roche and sought to hold them financially accountable. She filed a lawsuit against each, her psychiatrist for "malpractice in having prescribed excessive amounts of the drug Valium and then causing her to discontinue it abruptly," and Roche for "manufacturing and distributing Valium without ascertaining its dangers and without warning the public of its harmful side effects."[60]

Gordon was not the only American who blamed drug companies as well as doctors. Reporters, politicians, and activists—some identifying with an emergent "just-say-no" anti-drug philosophy, others emboldened by the feminist women's health and consumer movements—were also vocal. They blamed pharmaceutical firms for excessive drug promotion and for cultivating an "unnecessary" market for tranquilizers for everyday problems. At the 1979 hearings on the use and abuse of benzodiazepines, Senator Edward Kennedy told the audience that "The whole pitch appears to be to sell and market, to sell and market."[61] Companies had medicalized something that could not be fixed with a pill. These explanations resonated with Americans because, in part, they were true. Doctors had prescribed tranquilizers carelessly; companies had promoted them excessively. At the same, media reports erased the nuances of a complicated history of drug development and use. They did not, for instance, detail the commercial restraint that had characterized Miltown's release. Nor did they discuss the cultural and institutional enthusiasm for minor tranquilizers in the 1950s and early 1960s. Rarely did they mention that evidence of the drug's dependence liability had been well documented since 1961. Nor did they try to sort out the difficult question of what made Americans anxious in the first place.

Anxious Endings

Narratives of addiction by prescription did more than disturb audiences. Their frequent, public telling put tranquilizers themselves on trial and persuaded activists, politicians, pundits, and policymakers to support policies designed to contain the nation's penchant for pill popping.

Under the Controlled Substance Act of 1970, Valium and Librium became Schedule IV drugs, a category that included other prescription tranquilizers and stimulants. Five years later, the Justice Department adopted guidelines to encourage better medical oversight of tranquilizer use. The measure limited the number of refills a patient could get on a prescription of minor tranquilizers to five. It also set the expiration date for refills at six months. In 1980, the Food and Drug Administration took the unprecedented step of recommending that bottles of minor tranquilizer be labeled with the statement: "Anxiety or tension associated with the stress of everyday life usually does not require treatment with an anti-anxiety drug."[62]

In keeping with revised thinking about the hazards of tranquilizers, courts and juries began to award stiff fines and lengthy sentences to those caught harboring them illegally. In 1978, an Illinois guard, undertaking a routine pat-down among prospective visitors at a prison, discovered a woman with seven Valium pills in her jacket. Confronted with the evidence—unauthorized cargo at a penitentiary—the woman admitted her mistake. The admission was deemed irrelevant. The Circuit Court sentenced her to two to six years of imprisonment. The defendant appealed, arguing that the sentence was "improper and excessive." An Illinois appellate court ruled otherwise and upheld the original sentence.[63]

In Alabama, a man who accidentally dropped a matchbox containing "nine yellow medicine tablets" in the presence of a police officer met an equally severe fate. The man admitted that the drugs, acquired from his roommate, were Valium. Fifteen years earlier, the policeman might have let the matter go; sharing tranquilizers with friends and relatives was commonplace—something to scold a man for, perhaps, but nothing more. But this was 1979 and times had changed. The "yellow medicine tablets" were sent to the Alabama Department of Forensic Sciences where tests confirmed their chemical identity. The man was charged, tried, and sentenced to four years in prison for "unlawfully, willfully, and feloniously possess[ing] Diazepam."[64]

Those who had witnessed firsthand the evolving perceptions of tranquilizers wondered if their stigmatization was not doing more harm than good. The eminent psychiatrist Heinz Lehmann had expressed concern in 1960 that the popularization of minor tranquilizers such as meprobamate had led to "tremendous" abuse. Doctors were prescribing outside normal ranges and patients were having a hard time discontinuing their medicine. Decades later, Lehmann worried that the pendulum of political opinion had swung too far in the opposite direction. In the 1980s, Lehmann told reporters that patients had been hoodwinked by "sensational, horror stories that equate tranquilizer with addiction." Irresponsible media treatment had caused thousands of patients to forego medication and to suffer needlessly.[65]

Lehmann's perspicacious remarks point to the importance of understanding pharmaceutical drugs in social context. More than chemical compounds, drugs are social objects whose interpretation can shift dramatically over time, even when the chemical properties of the drugs in question vary little. The trials and tribulations of tranquilizers reveal as much about the society that passed judgment on them as it does the drugs' chemical powers and properties. If the culture and politics of 1950s America provided a particularly fertile ground for meprobamate, the more pessimistic mood of the 1970s and 1980s helped marginalize a group of drugs that were, according to most medical accounts, helping more than they hurt.

But this was not the end of the tranquilizer story. Taking drugs to relieve anxiety proved a hard habit to break. Although sales of Valium and Librium began to decline in the 1970s, sales of new-and-improved tranquilizers, such as the short-acting Xanax, introduced in 1981 for the newly created diagnosis of "panic disorder" soared. At the same time, pharmaceutical firms capitalized on the commercial vacuum created by the benzodiazepine backlash to launch Prozac and other SSRI antidepressants. By the late 1990s, these drugs were being repositioned as anti-anxiety agents too. Today, a full fifty years after Miltown revolutionized social and medical habits, drugs for anxiety are a billion-dollar business as Americans continue to search for ways to find peace in turbulent times.

N O T E S

The author would like to acknowledge the research assistance of Nathan Flis and especially Brian Pierce, and to thank Liz Watkins for her careful review of earlier drafts of this chapter.

1. NBC Evening News Broadcast, May 22, 1978, located at Television News Archives, Vanderbilt University, Nashville, Tennessee.

2. Howard La Fay, "All Wound Up? Here's a New Drug to Calm You Down," *Town Journal* (May 1956), 72; Susan Speaker, "From 'Happiness Pills' to 'National Nightmare': Changing Cultural Assessments of Minor Tranquilizers in America, 1955–1980," *Journal of the History of Medicine and Allied Sciences* 52 (1997): 338–76.

3. Thomas Whiteside, ""Onward and Upward with the Arts: Getting There First with Tranquility," *The New Yorker,* May 3, 1958, 99; Elizabeth McFadden, "Tension Busters," *Newark Sunday News,* May 19, 1957, 17.

4. *Use and Misuse of Benzodiazepines,* Hearing Before the Subcommittee on Health and Scientific Research of the Committee on Labor and Human Resources, United States Senate, Ninety-Sixth Congress, First Session on the Examination of the Use and Misuse of Valium, Librium, and other Minor Tranquilizers (GPO: Washington, D.C., 1979), 1.

5. Frank Berger, "My Biography," unpublished manuscript located in Frank Berger Collection, International Archives of Neuropsychopharmacology, Nashville University, Nashville, Tennessee; Andrea Tone, "Interview with Frank Berger, July 2003," Berger Collection, International Archives of Neuropsychopharmacology.

6. Berger, "My Biography," 7.

7. Whiteside, "Onward and Upward with the Arts," 113.

8. Frank Berger, "Anxiety and the Discovery of the Tranquilizers," in Frank Ayd and Barry Blackwell, eds., *Discoveries in Biological Psychiatry* (Philadelphia: Lippincott, 1970), 121.

9. Testimony of Henry Holt, *Administered Prices: Hearings Before the Subcommittee on Antitrust and Monopoly of the Committee on the Judiciary*, United States Senate, Eighty-Sixth Congress, pt. 16 (GPO: Washington, D.C., 1960), p. 9108; "The Social Chemistry of Pharmacological Discovery: The Miltown Story," *Social Pharmacology* 2 (1988): 190.

10. Carter Products Memo, October 21, 1957, Frank Berger scrapbooks, New York City, in the possession of Frank Berger; Address of Henry Hoyt before the New York Society of Security Analysts, June 9, 1958, Frank Berger scrapbooks; Whiteside, "Onward and Upward with the Arts," 114.

11. Whiteside, "Onward and Upward with the Arts," 112; Brian Pierce, "'No Substitute for a Martini': Frank M. Berger and the Tranquilization Process," M.A.

Thesis, McGill University, August 2005, 14–15; "The Social Chemistry of Pharmacological Discovery," 192.

12. John Swann, "FDA and the Practice of Pharmacy: Prescription Drug Regulation Before the Durham-Humphrey Amendment of 1951," *Pharmacy in History* 36 (2) (1994): 55–70; Peter Temin, *Taking Your Medicine: Drug Regulation in the United States* (Cambridge, MA: Harvard University Press, 1980), 50–53.

13. See, for instance, J. Maurice Rogers, "Drug Abuse—Just What the Doctor Ordered," *Psychology Today* (September 1971), p. 16.

14. Crosby quoted in Nathan Hale, "From Berggasse XIX to Central Park West: The Americanization of Psychoanalysis, 1919–1940," *Journal of the History of Behavioral Sciences* 14 (1978): 307.

15. Berger, "My Biography," 58–60; "The Social Chemistry of Pharmacological Discovery," 192.

16. Berger, "My Biography," 56–60.

17. "Tranquilizers—Successors to Aspirin?" *Chemical Week*, August 25, 1956, 18; La Fay, "All Wound Up?" 72.

18. Francis Bello, "The Tranquilizer Question," *Fortune* (May 1957), 162.

19. Donald Cooley, "The New Nerve Pills and Your Health," *Cosmopolitan* 157 (January 1956): 70–77.

20. Schmidt, "What you should Know about Those New 'Happiness Pills!'" *Uncensored!*, October 25, 1957, 37.

21. Kline quoted in "Soothing—But not for Drug Men," *Business Week*, March 10, 1956, 32.

22. Andrea Tone, "Heinz Lehmann: There at the Revolution," *Collegium Internationale Neuro-Psychopharmacologicum Bulletin* (March 2004): 16–17.

23. Lowell S. Selling, "Clinical Study of a New Tranquilizing Drug: Use of Miltown," *Journal of the American Medical Association* 157 (1955): 1594–96; Joseph C. Borrus, "Study of Effect of Miltown on Psychiatric States," *Journal of the American Medical Association* 157 (1955): 1596–98.

24. David Healy, *Let Them Eat Prozac: The Unhealthy Relationship between Depression and the Pharmaceutical Industry* (New York: New York University Press), chapter one.

25. Miltown ad *in American Journal of Psychiatry* 116 (July 1959).

26. Wallace Laboratories, *Miltown, The Tranquilizer with Muscle Relaxant Action: Physicians' Reference Manual, Fourth Edition* (Wallace Laboratories, 1958), National Library of Medicine.

27. Roy Porter, "'The Patient's View: Doing Medical History from Below," *Theory and Society* 14 (1985): 175–98.

28. Wallace Laboratories, *Miltown, The Tranquilizer with Muscle Relaxant Action: Physicians' Reference Manual, Fourth Edition*, p. 49.

29. Charles Grutzner, "Grave Peril Seen in Sleeping Pills," *New York Times*, December 16, 1951, 1.

30. Deborah Larned, "Do You Take Valium," *Ms. Magazine* 4 (1975), 27.

31. "Don't Give a Damn Pills," *Time*, February 27, 1956, 98–99.

32. La Fay, "All Wound Up?" 72.

33. Schmidt, "What you should Know about Those New 'Happiness' Pills," 60–62.

34. Jonathan Metzl, *Prozac on the Couch: Prescribing Gender in the Era of Wonder Drugs* (Durham: Duke University Press, 2003).

35. "'Miltown' Drug Output Raised," *Journal of Commerce*, May 8, 1956, 12; "Pill vs. worry—how goes the frantic quest for calm in frantic lives," *Newsweek*, May 21, 1956, 68–70; "Don't Give a Damn Pills," 68–70; "The Tranquilizer Question," 162–88.

36. David Sears Houston, "Hollywood's Latest Pill Kick: 'Don't-Give-A-Damn' Drugs," *Top Secret* (July 1956), 12.

37. Quoted in La Fay, "All Wound Up?" 72.

38. "Walter Winchell of New York," *Daily Mirror*, September 5, 1956, 10.

39. Jonathan Metzl, "Mother's Little Helper: The Crisis of Psychoanalysis and the Miltown Revolution," *Gender and History* 15 (August 2003): 228–55.

40. Irvin Cohen, "The Benzodiazepines," in Ayd and Blackwell, eds., *Discoveries in Biological Psychiatry*, 130.

41. "New Way to Calm a Cat," *Life*, April 18, 1960, 93–94; John Kinross-Wright, Irvin M. Cohen, and James A. Knight, "The Management of Neurotic and Psychotic States with Ro 5–0690 (Librium)," *Diseases of the Nervous System* 21 (March 1960): Suppl: 23–26.

42. Milton Moskowitz, "Librium: A Marketing Case History," *Drug and Cosmetic Industry* 87 (October 1960): 462.

43. Moskowitz, "Librium," 461.

44. Leo Hollister, Francis P. Motzenbecker, and Roger O. Degan, "Withdrawal Reactions from Chlordiazepoxide (Librium)," *Psychopharmacologia* 2 (1961): 63–68.

45. David Healy interview with Leo Hollister, "From Hypertension to Psychopharmacology—a Serendipitous Career," in Healy, ed., *The Psychopharmacologists 2* (London: Chapman & Hall, 1998), 225–26.

46. Hollister et al., "Withdrawal Reactions from Chlordiazepoxide (Librium), 63–4.

47. Note from Tom Ban to Andrea Tone, May 30, 2005, in author's possession.

48. John Pekkanen, "The Tranquilizer War: Controlling Librium and Valium," *The New Republic*, July 19, 1975, 17–19.

49. Edward Shorter, *A Historical Dictionary of Psychiatry* (New York: Oxford University Press, 2005), 41; Gilbert Cant, "Valiumania," *New York Times Magazine*, February 1, 1976, 34–41.

50. Andrea Tone, "Listening to the Past: History, Psychiatry, and Anxiety," *Canadian Journal of Psychiatry* 50 (June 2005): 378; "Valium Celebrates 40th, but not with a Bang," *Times Colonist*, July 21, 2003: D4.

51. Tone, "Listening to the Past," 378.

52. Shorter, *A Historical Dictionary of Psychiatry,* 42.

53. David Courtwright, *Forces of Habit: Drugs and the Making of the Modern World* (Cambridge, MA: Harvard University Press, 2001), 89; Pauline Maier, Merritt Roe Smith, Alexander Keyssar, and Daniel Kevles, *Inventing America: A History of the United States,* vol. 2 (New York: W.W. Norton, 2003), 958.

54. Erika Dyck, "Flashback: Psychiatric Experimentation with LSD in Historical Perspective," *Canadian Journal of Psychiatry* 50 (June 2005): 381–87; David Healy, *The Creation of Psychopharmacology* (Cambridge, MA: Harvard University Press, 2002), 193.

55. Leonard S. Brahen, "Housewife Drug Abuse," *Journal of Drug Education* 3 (Spring 1973): 13.

56. Carl D. Chambers and Dodi Schultz, "Women and Drugs: A Startling Journal Survey," *Ladies' Home Journal* (November 1971): 191.

57. "The Prisoner of Pills," *Newsweek,* April 24, 1978, 77; Penelope McMillan, "Women and Tranquilizers," *Ladies' Home Journal* (November 1976): 164–67.

58. Ruth Cooperstock, "Sex Differences in Psychotropic Drug Use," *Social Science Medicine* 12 (July 1978): 179–86; Ruth Cooperstock and Henry L. Lennard, "Some Social Meanings of Tranquilizer Use," *Sociology of Health and Illness* 14 (December 1979): 331–47; "Non-Working Wives Over 34 Are Biggest Users of Tranquilizers, Research Shows," n.d., Hoffman-La Roche Manufacturers Files, FDA History Office; Bart quoted in McMillan, "Women and Tranquilizers," 165.

59. J. Maurice Rogers, "Drug Abuse—Just What the Doctor Ordered," *Psychology Today* 5 (September 1971): 16–24; William Nolen, "Tired? Nervous? Here's a Pill," *McCall's* (April 1973): 16, 18, 22.

60. Barbara Gordon, *I'm Dancing as Fast as I Can* (New York: Harper and Row, 1979); *Barbara Gordon v. Roche Laboratories,* 456 N.Y.S. 2d 291, December 14, 1981.

61. Kennedy quoted in *Use and Misuse of Benzodiazepines.*

62. "Tightening the Lid on Legal Drugs," *Science News* 107 (June 14, 1975): 382; Pekkanen, "The Tranquilizer War," 18–19; ABC Evening News Broadcast, July 10, 1980, Television News Archives, Vanderbilt University.

63. *The People of the State of Illinois v. Leslie Audi,* 392 N.E. 2d 248, June 22, 1979.

64. *Rufus C. Cockrell v. State,* 392 So. 2d 541, October 7, 1980.

65. Tone, "Heinz Lehmann," 16–17.

Part III

7

STATINS

The Abnormal and the Pathological
Cholesterol, Statins, and the Threshold of Disease

Jeremy A. Greene

What is your cholesterol? Odds are about even that you can answer this question with a number, or at least a value. Your cholesterol is 198. Your cholesterol is normal. Your cholesterol is 250. Your low-density lipoprotein (LDL)-cholesterol fraction is above 130 milligrams per deciliter, or you have too much "bad" cholesterol, or, perhaps, you just "have cholesterol." Recent surveys indicate that between 50 and 75% of Americans over the age of 20 have had their cholesterol checked in the last five years.[1] Knowledge of one's cholesterol number has become for many adult Americans an essential act of self-surveillance, a window into one's inner health.

Over the last two decades of the twentieth century, a general consensus on the importance of cholesterol management has vaulted cholesterol-lowering agents from a minor therapeutic category to the leading class of prescription drugs sold in the world. This widespread enthusiasm for the detection and treatment of elevated blood cholesterol did not merely arise from the passive diffusion of a body of scientific knowledge through the general population. Rather, the present state of affairs is the product of a concerted public-private effort to make high blood cholesterol a priority for American physicians and consumers, involving to a significant extent the material and commercial attributes of the class of pharmaceuticals called the "statins." This chapter narrates the life and times of Merck,

Sharp and Dohme's *Mevacor* (lovastatin), the first and arguably the most influential statin in the American market, in relation to changing national guidelines on the treatment of high cholesterol.

The historical relationship between the statins and the treatment of high cholesterol demonstrates the hydraulics by which commercially funded clinical trials have come to drive the production of clinical guidelines and the standardization of clinical practice.[2] As definitions of normal and pathological have been remodeled around *Mevacor* and its fellow statins, "high cholesterol" has proven to be a remarkably expansive condition with highly mobile boundaries. In the wake of a series of expansive, privately funded clinical trials demonstrating preventive benefit at successively lower points along a gradient of risk, the population of those deemed to have high cholesterol and the corresponding market for statins have grown in proportion. This mutually reinforcing and expanding equation of disease and market recalls the more publicly discussed "diagnostic bracket creep" of depressive disorders into states previously defined as healthy, but with a twist. The relationship between statins and cholesterol demonstrates that the expansive trajectory of contemporary medical conditions is not limited to the psychiatric field, but is instead present at the very heart of somatic medicine.[3] As agents that reduce the *risk* of symptoms rather than relieving symptoms themselves, statins illustrate a logic of pharmaceutical prevention that became an increasingly important aspect of American medical and pharmaceutical practice in the late twentieth century.[4]

The double names of pharmaceuticals—for example, *Mevacor* (lovastatin)—often lead observers to regard these agents as having separable identities that navigate two distinct spheres: a generic, scientific domain in which drugs are discovered, tested, and rationally used, and a branded, marketing domain in which drugs are excessively advertised, sold, and subject to the taint of the profit motive. But this double play distracts us from the interconnected, polysemic fashion in which drugs and knowledge related to drugs are produced, circulated, and consumed.[5] Most knowledge accumulated around drugs relates in some way to its market identity, and marketing rationales have come to inform both the type of research conducted and the manner in which pharmaceutical knowledge influences therapeutic categories. *Mevacor's* clinical trials and product promotion walked a fine line between emphasizing elevated cholesterol as a legitimate disease with discrete pathology, while simultaneously deemphasizing the severity of that condition so that it might be understood as a common and familiar condition and thus reach the broadest possible market. As a prominent blockbuster drug in a period when the production of blockbusters was becoming increasingly important to the marketing priorities of drug firms, *Mevacor* captures the zeitgeist of the pharmaceutical industry of the 1980s and 1990s much as the sulfa drugs did in the 1930s and penicillin did in the 1940s.[6]

Now that *Mevacor* is off-patent, any study of its branding and marketing has become a historical project, and an interested historian can pursue a Freedom of Information Act (FOIA) request for documents surrounding *Mevacor's* launch. Such materials, read in conjunction with the clinical literature, industry trade literature, and the public records of the National Cholesterol Education Program, provide a sketch of how *Mevacor* was able to expand its own market through a series of interconnected clinical trials and evidence-based guidelines. This approach has its limitations; there is not space in this essay to properly flesh out the gaps that exist between guidelines and practice, or to provide an ethnographic sense of how consumers have learned to live with statins in terms often quite different from those intended by pharmaceutical manufacturers, regulators, or prescribers. Nonetheless, this applied study of marketing, research, and changes in clinical guidelines can provide a compelling illustration of the unitary economy of drug-knowledge and disease-knowledge in contemporary American medicine.

Twilight of a Drug

Mevacor's life as an active brand was cut short in the summer of 2000 by a federal tribunal in a suite of the massive Parklawn building in Rockville, Maryland, where most of the Food and Drug Administration (FDA) resides. True, *Mevacor* still exists today, in a ghostly sense: Merck plants continue to produce blue pills with "MSD" stamped into the side and the *Physician's Desk Reference* still carries an entry for the drug alongside other generic versions of lovastatin. But *Mevacor* inspires no large-scale clinical trials or advertisements in medical journals or popular magazines; its name graces neither pen nor desk pendant in the pharmaceutical sales representative's bag of gifts. Its patent lost, it has receded gracefully into the ranks of discount medicines and medicines deemed essential for care in developing nations but not essential to Merck's own portfolio.[7]

Perhaps the events in the Parklawn building that ended *Mevacor's* market life would best be described as a form of negative euthanasia, of a health care system refusing to provide life-support for an ailing organism. *Mevacor's* situation by June 2000 was already critical: the 20-year patent on lovastatin was due to expire the following year, Merck's second-entry statin, *Zocor*, had sucked away most of *Mevacor's* promotional budget and was already more popular among consumers and physicians, and generic manufacturers were developing plans to produce their own lovastatin and eat into the remaining market.[8] *Mevacor's* last chance for survival as a brand was to follow the lead of other faded blockbuster drugs and weather the switch from prescription (Rx) to over-the-counter (OTC) status.[9] Like the anti-ulcer agent *Tagamet*—the first prescription drug to be called a "blockbuster" and one of the first to successfully switch to OTC—*Mevacor* had developed a well-known brand name and widespread confidence in its safety and desirability as a consumer product.[10] The firm announced plans for the *Mevacor OTC* project in the spring of 1999, and submitted a formal petition for nonprescription status to the FDA in early June 2000.[11]

However, as many analysts pointed out, unlike the ulcer-blocking *Tagamet, Zantac,* and *Pepcid* and the pain-relieving *Tylenol, Advil,* and *Aleve,* the cholesterol-lowering *Mevacor OTC* would treat a condition unique for its lack of recognizable symptoms. All other drugs which had successfully switched to over-the-counter status treated conditions which patients could easily self-medicate. If someone had a headache and then took two *Advil,* she herself should be able to judge when the headache went away. If the symptom wasn't relieved by the *Advil,* the consumer would know to seek more formal health care. The same argument could be made for persistent stomach pains that didn't respond to *Tums,* or *Zantac,* or, indeed, for any other symptom unrelieved by available nonprescription drugs. As recently as 1997, the FDA had explicitly pronounced that over-the-counter drugs be used only for "self-recognizable conditions that are symptomatic, require treatment of short duration, and can be treated without the oversight and intervention of a health-care practitioner."[12] *Mevacor,* it was argued, could not possibly relieve symptoms if high cholesterol did not present any symptoms to relieve.

Or did it? By 2000, some advocates argued, cholesterol was so widely "felt" by the consumer populace that it could almost be considered a symptom. Consumers had been educated to feel ill if they had high cholesterol and to feel healthy if their cholesterol was low enough, and many studies began to document the subjective illness felt once an individual received a diagnosis of high cholesterol.[13] Reliable finger-prick cholesterol monitors were available in most pharmacies by 1999. Consumers with high cholesterol, concerned about their numbers, were already purchasing large numbers of nutriceuticals and other alternative medical products that claimed to lower cholesterol levels. Some of these products, such as red yeast rice supplements, contained naturally fermented lovastatin in levels comparable to the proposed *Mevacor OTC* but in an unregulated, nonstandardized form.[14] If consumers were already keeping tabs on their own cholesterol levels, Merck's representatives noted, and spending large sums of money on treating these numbers to their own satisfaction, then perhaps cholesterol had, in a sense, effectively become symptomatic to many Americans.

The FDA had appointed July 13, 2000 as a date to hear Merck's arguments for *Mevacor OTC,* but given the breadth of interest they also scheduled a public hearing a few weeks earlier to revisit the fundamental issues at stake in switching preventive medications from prescription to over-the-counter status.[15] When the public hearing was called to session in a Holiday Inn in Gaithersburg, Maryland, an odd collection of groups gathered to protest the switch. These included radical consumer groups such as the Ralph Nader-founded Public Citizen, physician groups such as the American College of Cardiologists, and Merck's chief competitor, Pfizer, which by 2000 had captured over 25% of the market with its new entry *Lipitor* and did not want to lose ground to a heavily marketed and familiar brand available on an over-the-counter basis. "We don't see how patients will be able to monitor their levels and treat to the right goal," the head of Pfizer's cardiovascular division was quoted in the newspapers, adding, "I think you really need a physician to check your levels and what your goals should be."[16]

It is not surprising that mainstream physicians' groups would oppose a switch that would remove their central role in the adjudication of risk. As Ed Frohlich, a spokesman for the American College of Cardiologists (ACC) noted:

> The ACC believes that the relief of symptoms should be an important requirement for OTC products. . . . [I]f relief requires a laboratory test, the consumer does not know whether he or she, in fact, are relieved. This is especially important for cardiovascular drugs which often can treat conditions which have no associated symptoms with which a consumer can assess the drug's efficacy.[17]

Frohlich's argument that the consumer could not gauge cholesterol as a symptom, however, was immediately critiqued on cross-examination by an FDA panelist who pointed out that such a distinction was overly simplistic:

> Ed, you draw a sort of bright line between treating symptoms and treating signs, I guess you could say, and one of the reasons is that a patient can't assess whether his sign has improved without some external help. However, in two conspicuous areas, cholesterol and blood pressure, you can go to your Giant Supermarket and get your latest blood pressure. I don't know how accurate those are, but you can do it, and there are or will be simple tests of cholesterol available. So a person who was taking an over-the-counter drug in order to modify those signs would, if they were interested in the first place, be able to see how they were doing, if they bothered.[18]

Merck advocates and several minority physician advocacy groups agreed that it was incidental whether the patient perceived a symptom bodily or through a mediating consumer technology. As the Association of Black Cardiologists and the Interamerican College of Physicans and Surgeons added, a more significant error lay in making the paternalistic and insulting assumption that only physicians could make sense of a number that the entire populace had been extensively educated to internalize.[19]

Merck lost its bid to extend *Mevacor's* branded life into the realm of consumer products.[20] On July 13, the FDA advisory panel voted eleven to one against *Mevacor OTC,* and the company, after a few last-minute attempts to gain further patent extensions, gave up on continued brand exclusivity for *Mevacor.*[21] Nonetheless, even if elevated cholesterol was not considered a condition symptomatic enough for consumers to self-medicate, the debate itself was symptomatic of the extent to which the abnormal and the pathological had become approximated in popular and medical conceptions of high cholesterol.

The Normal, the Abnormal, and the Pathological

Even as late as 1980, there was no popular or medical consensus on the value of treating high cholesterol on a preventive basis. In that year, the National Academy of Sciences issued a widely publicized report that reiterated the lack of convincing evidence that detecting and treating elevated cholesterol helped to prevent heart disease or death.[22] Although the Framingham Study had produced convincing epidemiological data as early as 1957 that linked high cholesterol to atherosclerotic heart disease, the benefit of interventions which lowered cholesterol remained a subject of debate for decades.[23] Unlike the treatment of high blood pressure, which progressed relatively swiftly from the evaluation of several relatively safe and effective medications in the 1950s and 1960s to the publication of the positive results of large-scale prevention trials in the early 1970s and the creation of nationwide detection, education, and guideline efforts, the treatment of high cholesterol inspired largely unsafe, ineffective, or unpalatable drug development throughout the 1950s, 1960s, and 1970s.[24] Parallel efforts to demonstrate the value of cholesterol-lowering diets also failed to demonstrate any benefit. It was not until 1984 that any large-scale randomized controlled trial could claim evidence of significant cardiovascular mortality benefit from reducing cholesterol levels; the publication of this trial would lead to the formation of a National Institutes of Health (NIH) Consensus Statement on the Value of Treating High Cholesterol and the initiation of a National Cholesterol Education Program (which worked to bring together professional organizations, state and federal public health bodies, and several other private and public stakeholders around a common promotional message) in 1985.[25]

One form of high cholesterol, however, had been clearly delineated as a disease state even before the word "cholesterol" entered the medical dictionary. The pathological condition known as xanthomatosis, first clinically described in 1851 and well known by the time the structure of cholesterol was identified in the early twentieth century, was characterized by cholesterol levels so high that small fatty tumors called xanthoma would become symptomatically evident on the skin and other regions of the body.[26] This was high cholesterol at its most self-evidently pathological; a lesion-based model of disease grounded by familial clustering that suggested single-gene Mendelian inheritance.[27] By the early 1960s, the genetic basis of metabolic error had become more central to the diagnosis of xanthomatosis than the original symptom of the xanthoma itself. In the 1963 edition of the popular *Cecil-Loeb* textbook, xanthomatosis was referred to as "essential hypercholesterolemia" (Latinate for a pathological elevation of blood cholesterol levels) and could be diagnosed by a combination of laboratory measurement and genealogy. The presence of fatty tumors was "typical" but no longer necessary for diagnosis, but a clear distinction was nonetheless preserved between the *disease* of essential hypercholesterolemia and the mere *chemical marker* of elevated serum cholesterol. [28] That the former should be treated with available therapeutic agents was never in question. The latter category, however—the mere measurement of blood cholesterol elevation—found no therapeutic consensus at that time.

Subsequent attempts to quantify the condition of high blood cholesterol reproduced the gap between the frankly pathological and the merely abnormal. Observations of the serum cholesterol levels of overtly xanthomatous patients allowed for the description of a numerical threshold of pathology. Because no such lesions existed in patients with serum cholesterol below 400mg/dL, a range of cholesterol levels above 400 could be a potentially pathological finding even in a symptomless individual. [29] This pathological threshold was joined, from the other side of the spectrum, by a normal threshold. Using data from life-insurance examinations and hospital laboratories, physicians, epidemiologists, and actuaries represented the distribution of serum cholesterol in the general population as a bell-shaped curve, its center around 195 mg/dL with known variance and standard deviation. According to the logic of standard deviation, "normal" could be bounded as the set of values within two standard deviations of the mean, a cutoff that defined the middle 95% of the population as normal and bound the upper and lower 5% as abnormal extremes. By such a calculation, medical textbooks in the 1950s and 1960s listed 130–260 mg/dl as a normal range for serum cholesterol and considered values above 300 mg/dl to be abnormally high.[30] Abnormally high cholesterol was not the same thing as an overtly pathological lipid disorder; rather, it constituted a shadowy third space between health and disease.

Calculating hypercholesterolemia on the basis of deviation from the mean reflected an interpretation of the biologically normal defined explicitly in terms of statistical norms. That the threshold itself was arbitrary was well understood, but it was also understood that this arbitrary distinction related to some real underlying boundary between pathology and physiology. Analogy was made to Claude Bernard's conception of the "renal threshold" in diabetes: just as the elevation of blood sugar over a certain level would "spill over" into the urine to produce the symptoms of diabetes, elevation of blood cholesterol past a certain level would accumulate in tissues to produce xanthomatosis and atherosclerosis.[31] In this rather democratic regime, the populace itself became the reference point for health; disease could be defined by demarcating the statistical deviant, the small group of people quantitatively distant from the core center of the population.[32] This scheme would be dominant throughout the 1960s and 1970s and into the early 1980s.[33]

However, for many advocates who believed that cholesterol was a central culprit in producing heart disease, the normal American way of life—particularly the cholesterol-rich "American diet"—became implicated in the causation of cardiovascular pathology. Noting that the twentieth-century American epidemic of coronary heart disease was correlated with the rise of the well-fed and underexercised American body, and claiming that heart disease represented a true disease of civilization, these cardiologists, nutritionists, and other public health activists argued that the bell curve of the American population should explicitly *not* be seen as the repository of healthy values.[34] In its place, epidemiological studies documented that non-American populations, with lower mean cholesterol values, experienced substantially lower cardiovascular mortality.[35] This observation would become formally incorporated into the National Institutes of Health Consensus Conference on Cholesterol and Atherosclerosis in 1985, a collective pronouncement of enthusiasts for cholesterol treatment which incorporated a subtle but fundamental redefinition of high blood cholesterol at the same time that it announced the creation of a federally funded program to fight it. The report of the Consensus Conference shifted the boundary between normal and abnormal in a step that retained normative statistical techniques while refusing to acknowledge that the mean of the American population represented the state of health:

> Often, an abnormally high level of a biologic substance is considered to be that level above which is found the upper 5% of the population (the 95th percentile). However, the use of this criterion in defining "normal" values for blood cholesterol levels in the United States is unreasonable; because, *in part at least, a large fraction of our population probably has too high a blood cholesterol level.* A review of available data suggests that levels above 200 to 230 mg/dL are associated with an increased risk of developing premature coronary heart disease. It is staggering to realize that this represents about 50% of the adult population of the United States.[36]

The Consensus Report detached the distribution of the normal population from the distribution of the American population and relocated the desirable mean leftward toward the distribution of an idealized "preindustrial" population. In this value-laden shift, the U.S. population was neatly transformed from being the arbiter of normality to constituting the locus of pathology itself. In place of a threshold defining the upper 2.5% threshold as abnormal (300mg/dL), the committee inserted a new threshold, 240mg/dL, that intentionally defined the upper 25% of the adult population as abnormal, while those with levels over 300mg/dL were now known as pathologically severe hypercholesterolemics.[37]

As quantitative definitions of normal cholesterol evolved, the qualitatively defined pathology of the disorders of lipid metabolism also shifted. What had been listed in textbooks as a single disease—"xanthomatosis" in the 1950s and "essential hypercholesterolemia" in the 1960s—had by 1970 bloomed into a verdant nosology of disorders of lipid metabolism. These were enumerated as Types I through V (based on the lipid subfractions involved) and subdivided into subtype (a) or (b) depending on whether they represented genetic or acquired disorders. The presence of the symptom—xanthoma—was now a morphological term incidental to diagnosis.[38] What counted instead, diagnostically, was the analysis of lipoprotein profiles—the *molecular* rather than the morphological, histological, or genealogical presentation of metabolic error.

Type II lipid disorders became a particularly important site of negotiation between the manifestly pathological and the merely abnormal, for Type II could roughly be translated to mean, "pathologically high LDL-cholesterol," with no other distinguishing abnormalities. This category was divided into two camps: Type II(a), genetically inherited high cholesterol, was also termed familial hypercholesterolemia, a highly penetrant mutation that clustered in families and fulfilled the requirements for a single-gene Mendelian defect. Type II(b), the acquired form, however, was considered a pattern rather than a disease, an inchoate diagnosis of exclusion.[39] The taxonomy of lipid disorders would undergo a further contortion that would significantly narrow the intellectual space between abnormal and pathological. The 1980 *Harrison's* entry on lipid disorders was authored by Michael Brown and Joseph Goldstein, who replaced the Types I through V system, with a new taxonomy based on molecular genetics.[40] Familial hypercholesterolemia (FH), formerly known as Type II(a), was now reclassified as a "LDL-receptor deficiency." The fuzzier category of lipid disorders for which single-gene mutations had not yet been found were shunted to the end of the chapter, and Type II(b) was reclassified as a "primary hyperlipoproteinemias of unknown etiology," known as polygenic hypercholesterolemia.

Polygenic hypercholesterolemia can be roughly glossed as a molecular geneticist's shorthand for "we don't know why the cholesterol is high."

Polygenic hypercholesterolemia was a dummy variable, a place-holder that allowed statistical abnormality to become more easily commensurate with taxonomies of pathology. As Brown and Goldstein note as much in their *Harrison's* entry, in a passage that came far closer to equating the abnormal with the pathological:

> By definition, 5 percent of individuals in the general population have LDL-cholesterol levels that exceed the 95th percentile and therefore have hypercholesterolemia. . . . On the average, among every 20 such hypercholesterolemic persons, one person has the heterozygous form of familial hypercholesterolemia, and two have multiple lipoprotein-type hyperlipidemia. The remaining 17 have a form of hypercholesterolemia, designated *polygenic hypercholesterolemia,* that owes its origin not to a single mutant gene but rather to a complex interaction of multiple genetic and environmental factors. Most of the factors that place an individual in the upper part of the bell-shaped curve for cholesterol levels are not known.[41]

If these changing taxonomies of lipid disorders confused the average practitioner, in the public eye the overlapping terms of Type II hyperlipidemia, familial hypercholesterolemia, and polygenic hypercholesterolemia were easily blurred. In practice, the individual clinician had a wide degree of play with which to explain what high cholesterol meant, if he chose to, in terms of disease. If your brother and sister were tested and found to have abnormally high cholesterol, or hypercholesterolemia, and so did you, didn't that sound like "familial hypercholesterolemia"? It is reasonable to believe such confusion extended beyond patients to practitioners. Ultimately, this slippage between the abnormal ("having high cholesterol") and the pathological ("having a disorder of lipid metabolism") would play into the broader goals of the National Cholesterol Education Program and other parties interested in mobilizing high cholesterol as a treatable condition or, ideally, a *disease.*

Market Expansion and Disease Expansion

This slippage between high cholesterol as distinct pathology and high cholesterol as quantitative variation was also ideal from a marketing perspective, and *Mevacor's* marketing team—well underway with product development by 1985—paid close attention to both the shifting textbook definitions of lipid disorders and the National Cholesterol Education Program's activism regarding the lowering of the numerical threshold of normality.[42] Although Merck, Sharp and Dohme's marketing division predicted that launching *Mevacor* would be "its largest effort to date," *Mevacor's* marketers had to be very careful in their initial promotion of its drug.[43] In order to be approved by the FDA, every drug requires a therapeutic indication and ideally that indication should reflect an identifiably pathological state. To that end, the initial trials and new drug application (NDA) submitted for *Mevacor* carefully limited their claims to the disease of familial hypercholesterolemia, rather than for the treatment of elevated blood cholesterol levels in general.[44] This represented the first stage— legitimate market penetration—of what Merck's senior director of marketing planning, Grey Warner, explained to Merck employees and shareholders as a two-pronged marketing strategy: first to gain awareness of *Mevacor's* value in treating clearly diseased patients, and then to expand cholesterol awareness in the more general population.[45]

To assure the first goal—that *Mevacor* was an effective drug that treated an unambiguously pathological condition—Merck conducted the initial clinical trials in severely hypercholesterolemic patients with the single-gene diagnosis of familial hypercholesterolemia (FH).[46] *Mevacor's* product manager, Dr. Jonathan Tobert, emphasized the severity of this condition as justification for the use of a novel experimental therapy, citing several lipid-disorder specialists who worked with gravely hypercholesterolemic FH patients. "[T]hese clinicians," he noted, "said *Mevacor* might prove the only chance these patients had to lower life-threatening cholesterol levels."[47] Merck's director of cholesterol research, Alfred Alberts, recalled talking with a Merck clinician in early 1987:

> He was very anxious and he said all the cardiologists and all the primary care physicians down there want the drug, and I said, "Well, that's a little different than what our marketing people tell us that these are the toughest people to convince," and he had a very interesting retort to this, he . . . likened the disease—the severe form of hypercholesterolemia—to AIDS, in this regard: that people . . . were going to die in three, four years, unless you could do something drastic for them.[48]

Comparison of the AIDS patient and the *Mevacor* patient was, in 1987, neither an accidental nor a trifling statement, and calls attention to the need, in the early clinical testing of a potentially risky novel therapeutic agent, to limit clinical trials only to subjects who were sick enough to merit the risk.

Attention to this small population (roughly 400,000 in the U.S.) of severely affected hypercholesterolemics was carefully balanced with the second goal of expanding *Mevacor's* marketing toward the one out of every four American adults estimated to have abnormal cholesterol values above the NCEP threshold of 240 mg/dl. As early results of *Mevacor* in FH patients showed promising levels of cholesterol reduction with a good safety profile, Merck was able to organize a few trials testing the drug in patients who had cholesterol levels greater than 300 mg/dl but lacked other affected family members—or monogenetic mechanisms—required for diagnosis of the familial disorder.[49] The NDA that Merck eventually submitted on November 1986 was based on 750 of these mono- and poly-genetically severe hypercholesterolemic patients. When FDA approval of *Mevacor* was announced in September 1987, both indications were included.[50]

This dual indication allowed for a convenient blurriness in the terms available to promote the drug at the time of its launch. Merck's press release announcing the launch of *Mevacor* also listed names and phone numbers of Merck-funded researchers who were available to comment on the significance of the drug. These individuals present striking examples of the academic-industrial dual citizenship that had increasingly come to define a successful career in academic medicine by 1987. The *New York Times*, for example, used the press release to contact Antonio Gotto, who was simultaneously a professor at Baylor University, active as a *Mevacor* clinical trials researcher and listed as a Merck spokesman, head of an NIH-funded Lipid Research Center, a planner within the National Cholesterol Education Program (NCEP), and the president of the American Heart Association.[51] Even as he announced that *Mevacor* had been fully tested and FDA approved only for severe lipid disorders with cholesterol levels over 300 mg/dl, Gotto was able to simultaneously suggest that he himself would prescribe the drug for all adults over 40 whose cholesterol levels were over 260 mg/dL "if they could not reduce their cholesterol by other means."[52] The slippage here between the drug's formal indication for severe patients and Gotto's personal recommendation for more widespread "moderate" usage was subtle but definite. Moreover, because Gotto could claim that he was merely stating his own clinical opinion, neither Merck nor the NCEP could be held responsible for his comments, regardless of his strong ties to both organizations. Through such powerful intermediary figures as the president of the

American Heart Association, broad off-label usage of *Mevacor* could be advocated without negative consequence.

Nor were Gotto's multiple ties to Merck and the National Cholesterol Education Program unusual. Nobel laureates Brown and Goldstein performed basic science research closely related to *Mevacor's* development and played prominent roles in the NCEP. The pair received the Nobel Prize for their cholesterol work in 1985, the same year in which the NIH Consensus Report was published and the National Cholesterol Education Program began. The surrounding publicity both added to the aura of scientific imminence surrounding cholesterol as a public health threat and worked to blur the boundaries between those areas in which high cholesterol was clearly related to a pathological process such as LDL-receptor deficiency and those areas in which high cholesterol was "polygenic," or merely abnormally elevated. Brown and Goldstein's colleague at the University of Texas, Scott Grundy, was the principal investigator in *Mevacor's* initial clinical trials. Gotto, Grundy, Brown, and Goldstein were also involved with a Merck-funded lobbying group, Citizens for Public Action on Cholesterol, on whose medical advisory panel all three belonged.[53] Mike Gorman, the veteran Washington health lobbyist and close colleague of Mary Lasker who directed Citizens for Public Action on Cholesterol during its brief career from 1985 to 1987, had a lengthy professional relationship with Merck marketers and liaisons Grey Warner and Russ Durbin, as well as a seat in the NCEP's central committee. Without any direct participation of Merck in the National Cholesterol Education Program, *Mevacor's* promotion would subsequently be tightly bound up with NCEP promotion.

Regardless of how tight this coterie of advocates may have been, however, the nascent system of cholesterol treatment guidelines required a continuous and trusted flow of new data in order to perpetuate itself in clinical practice. Merck, for its part, understood that large-scale clinical trials data in patients would be necessary to achieve the larger potential market of Americans with cholesterol levels that the national guidelines deemed abnormal. Long before the FDA approval of *Mevacor* was announced, postmarketing plans were already underway to seek broader therapeutic indications that could be applied more generally to this larger population. A few months before *Mevacor's* launch date, product manager Jonathan Tobert announced to the rest of Merck that physician comfort and familiarity with the drug would soon be bolstered by "a large-scale Phase V study just getting underway that will eventually involve over 7,000 patients."[54]

The Phase V study was a Merck-specific term for a type of postmarketing clinical trial that was gaining prominence in the therapeutic landscape of the late 1980s. The clinical trials sequence codified in the wake of the 1962 Kefauver-Harris Act had delineated four phases of clinical trials research: small Phase I trials emphasized tolerability in healthy subjects, larger Phase II and III trials emphasized efficacy and dose-response in patients, and Phase IV trials investigated the long-term safety and efficacy of a compound after FDA approval. What Tobert called Phase V studies (and other firms referred to as specialized Phase IV trials) were a set of expensive, large-scale, long-term trials conducted with the aim of developing additional therapeutic indications for an already-approved drug or broadening the terms of an existing indication. In other words, these were trials of market expansion. To get a sense of the role of these trials in the changing fiscal priorities of late-1980s pharmaceutical firms, one need only compare the sum total of *Mevacor* subjects Merck studied to obtain FDA approval—750 in all—to the size of just one of the many Phase V *Mevacor* trials Merck would come to support—listed at 7,000 subjects and growing when Tobert announced it in 1987.

Unlike *Mevacor's* earlier clinical trials in clearly pathological conditions such as familial hypercholesterolemia, these postmarketing studies explicitly staked out a role for *Mevacor* in the treatment of abnormally elevated cholesterol, using the population thresholds set out within the NCEP guidelines. Merck's "first Phase V" trial of *Mevacor* was called EXCEL, shorthand for "EXtended Clinical Evaluation of Lovastatin."[55] In addition to being the first study to use the national treatment guidelines as a guide to enrolling research subjects, the EXCEL trial was also unique in applying the national guidelines goals as its target endpoints.[56] Publication of the EXCEL trial validated *Mevacor,* and it also helped to validate the NIH Consensus Statement and the NCEP's guidelines that many clinicians had critiqued as a false consensus of arbitrary thresholds.[57] The overwhelming majority of subjects taking *Mevacor* were able to achieve the goal LDL-cholesterol levels set by the NCEP.[58] By generating data using the NCEP's numerical thresholds, the EXCEL trial helped to buttress them with a concrete empirical basis. In the wake of the trial, a direct flow linked the NCEP guidelines to *Mevacor* prescription in patients with abnormally high cholesterol.[59] The flow from therapeutic guideline to *Mevacor* prescription became even more pronounced two years later, when revised national guidelines from NCEP reclassified *Mevacor* as a first-line agent.[60]

However, by the time the revised guidelines were released, another important shift had altered the dynamics between trial and guideline even more drastically: *Mevacor* was no longer the only statin on the market. The 1991 launch of Bristol-Myers Squibb's *Pravachol* (pravastatin), the second statin available in the United States, and the near-simultaneous release of Merck's own second-entry *Zocor* (simvastatin), transformed the landscape of cholesterol clinical trials from an ambitiously expanding monopoly into a fiercely competitive arena with a billion-dollar market already at stake.[61] The resulting outpouring of competitive trials— designed not only to expand the market but also to wrest it away from direct competitors—gave the commercial clinical trial a more pivotal role in marketing and development and an increasing scale of funding and impact.

Trials, Indications, and Guidelines in the Competitive Marketplace

Where EXCEL demonstrates how a commercial clinical trial could validate and concretize guidelines into more substantive forms of clinical knowledge, the second generation of large-scale statin trials came to exert a formative influence on the guidelines themselves. Indeed, the statin trials of the 1990s provide a uniquely dramatic illustration of the central role which industry-funded trials have assumed in the economy of clinical knowledge production. Although EXCEL was sponsored by Merck and conducted in academic medical centers such as the University of Kansas, Louisiana State University, and the Baylor College of Medicine, the responsibility for the execution of the study was contracted to a small company in Research Triangle Park called Clinical Research International.[62] The company—an early contract research organization or CRO—was part of a nascent industry growing in this North Carolina academic-industrial suburb that marketed clinical trials services to pharmaceutical companies.

Though the CRO industry had its origins in the late 1970s in a series of small statistical and regulatory consulting groups, the growth in size and number of industry-sponsored clinical trials had, by the 1990s, led to a wave of growth and consolidation in the field. The result was an industry of transnational corporations that could, for a price, organize all aspects of a clinical trial for a drug company: from trial manufacturing of pills, to producing the human subjects who would take them, to tabulating the results, and easing passage through regulatory bodies.[63] By the end of the 1990s, 25% of drug development budgets were outsourced to CROs, and the industry had doubled to a $6.2 billion market worldwide with the top 20 companies making revenues of $50 million or more.[64]

The rapid rise of the CRO documents the increased demand for large postmarketing clinical trials and the augmented scale such trials required to show the benefits of pharmaceutical agents in subtle and/or asymptomatic conditions like elevated blood cholesterol. Whereas in the 1970s, most truly large-scale postmarketing trials had been funded by large bureaucracies such as the NHLBI and the World Health Organization, by the 1990s, every company with a statin interested in expanding market share needed its own version of EXCEL. At the beginning of the decade, nearly all industry funds for clinical trials were channeled through academic medical centers; by 1998, this figure was cut in half.[65] As the founder of one of the first CROs noted, the growing difference between "academic science" and "FDA science" had made academic medicine an increasingly inefficient partner for a pharmaceutical development.[66] Over the course of the 1990s, the clinical trial had become an industry unto itself.

In this context, it is not surprising that when Bristol-Myers Squibb launched its own statin *Pravachol* in 1991, with the intent of redirecting as much of *Mevacor's* $1 billion annual market as it could manage, the drug's developers concluded that the mere safety-and-efficacy demonstration required for FDA approval would not suffice to obtain the prescription and sales figures the company desired. Instead, *Pravachol* would enter the market fully equipped with a variety of large-scale, long-term clinical trials actively investigating broader secondary and primary prevention outcomes that BMS hoped to link exclusively to its product. Marketers at Merck, in turn, suspected that they could deflect *Pravachol's* trajectory from wounding *Mevacor's* market share by demonstrating that *Mevacor* was the more potent lipid-lowerer.[67] The bodies of clinical trials subjects soon became the battleground of a brand by 2000, at least 35 separate competitive trials comparing the cholesterol-lowering abilities of two or more statins had been published, with industry sponsorship central to all but three.[68]

Nevertheless, head-to-head trials were potentially damaging for either agent involved and could easily backfire. Marketers quickly realized the heightened benefit of trials which worked to expand the total market in a way that delivered particular advantage to their own products without risk of accidentally proving their own product inferior to the competing brand. By the late 1980s, the most promising avenue for the expansion of the statin was the field of secondary prevention. Whereas *primary prevention* involved the difficult task of motivating a healthy and symptom-free population toward taking a pill, *secondary prevention* concerned a population that had already suffered a heart attack or an ischemic event diagnostic of coronary heart disease—in other words, a population that already saw itself as diseased and uniquely motivated to prevent further heart disease. Since patients with known heart disease were at a much higher risk for further cardiac events, they also represented a population more poised to accept and benefit from cholesterol-lowering drug therapy. This expanding sequence of prevention trials linking the statins to broader categories of previously untreated patients is one example of a more general trend in clinical trials sequences also found in the preventive pharmacy of osteoporosis, breast cancer, and Alzheimer's disease.

Secondary Prevention Trials

Although the thresholds of normal, abnormal, and pathological cholesterol remained constant between the first and second national guidelines for cholesterol treatment, a secondary prevention population of coronary heart disease (CHD) patients was "carved out" of the new guidelines and placed under more aggressive treatment on a basis of lipoprotein fractions, even if their total cholesterol levels were normal.[69] By the time that the revised guidelines were published in 1993, Merck and Bristol-Myers Squibb were already underway with dueling large-scale, long-term studies in a race to market the first—and therefore, for a crucial marketing window, the only drug with demonstrated benefit in the secondary prevention of recurrent heart disease.[70]

The advantage this time was Merck's, but not *Mevacor's*. The first publicized results of a secondary prevention trial featured Merck's newer product *Zocor* (simvastatin), which, though more potent than *Mevacor*, was less well known and not yet selling as well.[71] This trial, the Scandinavian Simvastatin Survival Study (popularly known as "4S"), was initiated four years prior to *Zocor's* FDA approval and involved nearly 4,500 patients and 500 clinician-researchers in five different countries.[72] When the trial's results were made public in November 1994, the group of subjects receiving *Zocor* had experienced roughly one-third less cardiovascular mortality than that seen in the control group. The results were promoted as international news, the headline "Cholesterol Drugs Found to Save Lives" made the front page of the *New York Times*, and the Nobel laureate Michael Brown was quoted in national newspapers, describing the results as "pivotal" and "absolutely astonishing": the first prevention trial that had satisfactorily documented that lowering cholesterol could actually reduce mortality.[73]

Prior to the 4S trial, all cholesterol-lowering drugs had been required to include in their labels and advertisements a disclaimer that there was "no definite link" between lowering cholesterol levels and lowering the rate of developing a heart attack; but in the wake of 4S, the FDA formally approved a new indication allowing *Zocor*—and only *Zocor*—to claim the ability to reduce mortality from heart attacks and prevent recurrent heart disease. This new promotional possibility gave *Zocor* room to grow its market, and surveys indicated that as much as 75% of secondary-prevention eligible populations were not yet being treated.[74] *Pravachol's* own major secondary prevention study, entitled Cholesterol and Recurrent Events (CARE), would not be published for another two years, and although *Pravachol* would ultimately receive an expanded secondary-prevention indication, the lag in publication had given its competitor a two-year competitive advantage.[75]

The reception of the *Zocor* and *Pravachol* secondary prevention trials also marked a turning point in the relationship between commercial clinical trials and clinical practice guidelines. The public impact directly led to a further revision to the NCEP guidelines regarding the role of statins in the care of the CHD patient. In 1997, a supplement to the guidelines was published to incorporate the 4S and CARE data, further lowering the treatment threshold for pharmacological therapy in CHD patients and creating several million additional candidates for statin therapy. The revised guidelines explicitly described populations of patients in terms of the commercial trials—and hence, the pharmaceutical agents—that had rendered them therapeutic candidates. As the guidelines noted, there were a projected 3.5 million "4S-like patients" to whom, in the aftermath of the trial, it was now unethical *not* to offer *Zocor*.[76] In a subtle inversion of earlier trials like EXCEL, commercial postmarketing trials of the mid-1990s were no longer reacting to national guidelines, but were instead beginning to drive them.

Primary Prevention Trials

The success of secondary prevention trials in broadening the potential market for statins, through expanded therapeutic indications and broader national guidelines, only enhanced the stakes for similar trials in the much larger market of primary prevention among otherwise healthy individuals. Once again, Merck and Bristol-Myers Squibb developed evenly matched, $50 million dollar prevention trials pitted against one another, each enrolling thousands of clinical trial subjects over several years. That the desired data would, in fact, emerge from these trials was to many industry analysts a tacit assumption; the question they focused upon was which drug, and which pharmaceutical firm, would be favored with the first positive data and the promotional advantage that came with earlier results. As a trade journal noted in the early fall of 1995:

> Once the results are published (and assuming they are positive) the fight for market share will begin in earnest. With statins influencing both primary and secondary prevention, the total base will be huge. Jockeying for position will be fluvastatin (Sandoz), simvastatin and lovastatin (MSD) and pravastatin (Bristol-Myers Squibb), and poised in the wings atorvastatin (Parke-Davis).[77]

Bristol-Myers was ultimately the fastest in this race for trial results; in a public-relations coup the results of *Pravachol's* primary prevention trial would be simultaneously published in the *New England Journal of Medicine,* announced at the American Heart Association meetings, and reported in a *Wall Street Journal* article, declaring *Pravachol's* first significant victory against *Mevacor.*[78] The front-page splash in the *New York Times* (even larger than the analogous headline which *Zocor's* 4S study had received the year before) declared that the publication of *Pravachol's* West of Scotland Coronary Prevention Study (WOSCOPS) had shown "for the first time that one of the potent new cholesterol-lowering drugs can prevent heart attacks and coronary deaths in apparently healthy men."[79] In a special editorial accompanying the WOSCOPS study, Torje Pedersen—the principal investigator of the *Zocor* 4S study—noted that the West of Scotland study had completed the final link in the logic of cholesterol lowering, offering hard outcomes data demonstrating the benefits of lowering high cholesterol levels in an otherwise healthy population.[80]

The benefits of *Pravachol's* Scottish study extended well beyond immediate publicity. Within eight months, their new labeling submission had been approved by the FDA for a primary prevention indication.[81] Under the new provisions, *Pravachol* alone was licensed to claim the ability to prevent heart attacks in people with elevated cholesterol levels but no other risks of heart disease. This led to a massive direct-to-consumer (DTC) print advertising campaign in which Bristol-Myers Squibb took out two-page spreads in the front section of the *New York Times* and other newspapers and magazines which called *Pravachol* "the first and only cholesterol drug of its kind proven to help prevent first heart attacks."[82]

WOSCOPS had caught Merck unprepared, as their own primary prevention study had required more time to develop a finding (eight years as opposed to five) and was not published until 1998. This gave *Pravachol* more than two years on the market as the only statin that could directly promote therapeutic claims for primary prevention, and when Merck's own study was eventually published, it received far less publicity. The editorial in the *Journal of the American Medical Association* accompanying Merck's publication initially referred to it as "yet another randomized, placebo-controlled, clinical trial of a 3-hydroxy-3-methylglutaryl coenzyme A (HMG-CoA) reductase inhibitor, in this case lovastatin, as a means to prevent atherosclerotic coronary artery disease."[83] Nonetheless, *Mevacor's* major primary prevention study, the Air Force/Texas Coronary Atherosclerosis Prevention Study (AFCAPS/TexCAPS) would prove a swan song of sorts for the aging drug. After eight years, *Mevacor*-consuming subjects experienced more than one-third fewer cardiac events than their placebo-consuming counterparts. What made the study results particularly impressive, however, was that most of the overtly healthy men and women enrolled in the trial had cholesterol measurements within the contemporary boundaries of normal.[84] Fewer than one out of five of these subjects would have previously qualified for statin therapy under existing treatment guidelines.[85] The following year *Mevacor* received an FDA indication which industry analysts heralded as "the first approval to market a statin in a generally healthy population."[86]

By the time the third revision of the national cholesterol treatment guidelines were published in 2001, there were nearly thirty actively ongoing major preventive trials, only one of which was not industry-supported.[87] Between the second and third guidelines, LDL-cholesterol ("bad cholesterol") thresholds multiplied from a single line to a graded spectrum, including ideal, above optimal, borderline high, high, and very high.

Following the *Mevacor* primary prevention trials, the line separating normal from abnormal HDL-cholesterol ("good cholesterol") levels shifted upward from 35mg/dL to 40 mg/dL, further increasing the ranks of the treatable. The new guidelines also denoted several qualifying conditions, such as diabetes, and a calculated risk score from a separate worksheet, which placed a person *without* coronary heart disease into a "coronary heart disease-equivalent" category that merited the aggressive lipid-lowering therapy of secondary prevention.[88]

Taken in sum, the 2001 guidelines nearly tripled the proportion of the U.S. population that would be eligible for lipid-lowering therapy to a market of 36 million people, and commercial clinical trials had played a key role in driving that expansion. When confronted with the observation that most of the guideline committee members had financial ties to the companies who produced statin drugs, the committee chair, Scott Grundy, responded that it was impossible by the end of the twentieth century to find any medical expert who did not have strong industry ties. "You can have the experts involved," he noted, "or you could have people who are purists and impartial judges, but you don't have the expertise."[89] By focusing on the possibility of illicit influence of the pharmaceutical industry, however, Grundy's critic missed the more significant role of *licit* influence by which pharmaceutical manufacturers have used clinical research and clinical researchers to influence the NCEP guideline process. Even without the implication of any undue corporate influence in the guideline-setting process, it is evident that the commercial clinical trial had proven tremendously successful at expanding the role of statins to first fill the categories of NCEP guidelines and then exert an outward pressure on those categories, expanding the population of the treatable as thresholds for drug therapy decreased.

Defining the Bottom Line

The success of the statin trials in lowering treatment thresholds led many to wonder openly if there was indeed any limit to how far those boundaries could be pushed. As one investigator noted at the June 2000 *Mevacor OTC* hearings, the curve of benefit from prevention trials at successively lower and lower degrees of risk implied a potentially infinite extension of *Mevacor's* utility. Approving the statins for over-the-counter use was merely "the next logical step" in the progress of risk-reduction through cholesterol lowering.[90]

At stake in the *Mevacor OTC* hearings was not only the principle of unmediated access to the drugs but also the designation of a new population of patients not currently understood to fall within the NCEP guidelines for pharmacotherapy. Using evidence from the *Mevacor* primary prevention study, Merck sought to market *Mevacor OTC* to a population of proto-patients "with mild to moderately elevated cholesterol whose conditions are not severe enough to warrant prescription medicine." Based on its own primary prevention trials showing benefit in patients with borderline levels of cholesterol (200–240mg/dL), and borderline levels of LDL-cholesterol (130–190 mg/dL), Merck argued that the data on *Mevacor* justified not only the over-the-counter marketing of the drug but also a further redefinition of who had treatable cholesterol.[91] Jerome Cohen of St. Louis University, speaking for Merck, clarified this position with a discussion of the arbitrary line dividing normal from pathological when risk follows a graded continuum:

> When I began medical school, in fact, 300 was often called normal, and we've seen it drop to 280 and 250. Two-fifty offers twice the risk as 200, but let us look at 200, which is now considered so-called desirable. Do you want a level of 200? The answer, I would hope, when you know the data, is no, an ideal level which I would define as optimal levels of cholesterol is shown there at 150 milligrams per deciliter, which minimizes your risk for death from vascular disease, and we're moving in that direction. . . . What you can see is the preponderance of cholesterol levels from which coronary disease eventually arises is in this so-called mild elevation of cholesterol range. That's where the action is. That's where the majority of people are. That's the group that's often dismissed by physicians and say, "Well, our cholesterol is a little high, 210, 220." It's almost normal; it's almost average. Well, the average person in this country dies from coronary heart disease, and so you don't want to have an average level. You want to have an optimal level. Remember that if nothing more.[92]

By Merck's estimates, the treatment of this abnormal "borderline" population would bring an additional 30% of the U.S. population into the potential market for *Mevacor*.[93] Further expansion prospects also seemed bright, as national health survey data available at the time of Cohen's talk indicated that, were his ideal cholesterol level of 150mg/dl to be accepted, approximately 90% of the U.S. population would be defined to have higher than ideal cholesterol.[94] This is a dramatic inversion of the logic prevalent in the 1950s and 1960s which had defined normal precisely in terms of the cholesterol level represented by 95% of the population.

If a patient's blood pressure was lowered below a certain level, physicians knew that one risked entering a clearly pathological state of hypotension which was more acutely dangerous than hypertension itself. Serum cholesterol behaved differently. Even though rare metabolic disorders such as Gaucher's disease and Niemann-Pick syndrome have been associated with lower total and LDL cholesterol, the level at which excessively low cholesterol clearly causes harm to the organism has been much harder to define. Literature describing epidemiological linkages between low serum blood cholesterol and mortality from cancer and other noncardiovascular causes has receded as the conclusive evidence of mortality benefit from the series of statin prevention trials continues to mount.[95] And LDL-cholesterol seemed to have no bottom value at all; in 2003, the widely reported PROVE-IT trial comparing *Lipitor* and *Pravachol* greatly bolstered clinical consensus that there is little to risk, and much to gain, from reducing one's LDL-cholesterol levels to lower and yet lower levels.[96]

By the end of the 1990s, it had become a commonplace occurrence for cardiologists to suggest, only half-jokingly, that statins should be included along with fluoride as a general additive to the nation's drinking water supply. From the pharmaceutical industry's perspective, lowering the threshold for treatment represented a "win-win" arrangement between private industry and public health. The lower the threshold, the larger the market, the healthier the pharmaceutical economy. The lower the threshold, the less the mortal and economic costs to the nation from heart disease and stroke. Who could argue against such a convergence of benefit?

Critics of broader pharmaceutical treatment of cholesterol had found their space of argumentation within the medical literature steadily diminished. The career of Michael Oliver, perhaps the most prominent and well-respected critic of the National Cholesterol Education Program, illustrates particularly well the narrowing space of possible dissent. After critiquing the 1985 Consensus Conference as a "Nonsensus Conference," Oliver could still firmly claim in 1988 that there was no evidence that reducing cholesterol reduced mortality. The emergence of *Mevacor* and the NCEP did not in and of themselves silence Oliver's critical voice; indeed, in the early 1990s, Oliver penned several influential reviews concerning the possible dangers of cholesterol-lowering interventions. But in the wake of the statin trials of the 1990s, such claims were no longer publishable in mainstream medical journals. Though occasional critiques of cholesterol "dogma" continued to appear, by the end of the twentieth century they had been effectively marginalized, no longer appearing in refereed journals or professionally sanctioned monograph series, and largely reduced to a conspiratorial feature of the far left press.[97] With no absolute physiological grounds for opposing the progressive lowering of the threshold of treatability, those opposed to widespread statin consumption turned instead to economic and moral arguments.[98] The critique of widespread statin usage today remains split on the classical divide of ethical argumentation between utilitarian calculations of efficient use of resources and ethical arguments based on first principles. The resulting field is fragmented and unable to mount a unified opposition to the recursively empirical rolling back of treatment thresholds.

Utilitarian arguments against widespread statin use tend to founder on questions of metrics. In 1989, two U.S. and Canadian task forces independently suggested that applying NCEP guidelines would cost anywhere between $3 and $14 billion a year, with unclear benefit.[99] These early cost-effectiveness analyses of cholesterol treatment were initially deployed as a critical methodology used by central planners to evaluate whether costly interventions served in the best interest of the population as a whole. Both groups proposed that metrics like the "cost per life-year saved" and the "number of patients needed to treat" in order to prevent one death could serve as exchangeable metrics to compare interventions in a more pragmatic sense than the condition-specific concepts of efficacy and safety.[100]

Cost-effectiveness studies of cholesterol guidelines in the late 1980s calculated the costs of the NCEP guidelines to range from $32,000 to $606,000 per life-year saved, depending on age and risk profile of the population studied, with an average figure around $150,000 per life-year saved through drug therapy.[101] Based on such a high outlay per positive outcome, these critical analyses argued that screening the entire population was far from cost-effective. In a zero-sum economy, screening and cholesterol-lowering pharmacotherapy distracted from funds that could be more efficiently directed toward other cardiovascular preventive efforts with lower price tags, such as drug therapy for mild-moderate hypertension or even the cost of educating 10 year olds to reduce cholesterol.[102]

Planners of the national cholesterol guidelines also recognized the importance of the new logic of cost-effectiveness and incorporated cost-effectiveness of cholesterol-lowering therapy in the system of the revised (1993) guidelines.[103] Echoing a prevailing trend in health economics, the NCEP declared that spending less than the per capita output of the U.S.

economy (at that time, less than $20,000), per quality-adjusted life-year (QUALY) saved was "highly cost-effective," that interventions ranging from $20,000 to $50,000 a QUALY—such as cervical cancer screening, and hemodialysis—represented an "acceptable" range, while a cost of $50,000 to $100,000 "raised questions," of cost-inefficiency and any intervention with more than $100,000 per QUALY represented an "excessive expense and inappropriate utilization of resources."[104] By the NCEP's calculations, cholesterol-lowering in secondary prevention was clearly worthwhile—with a QUALY price-tag well below $20,000. Primary prevention in high-risk patients, with a QUALY price of $17,000 to $42,000, was also well within the acceptable range of cost-effectiveness. But general primary prevention with drug therapy in all categories of hypercholesterolemia, the NCEP noted, ran from $90,000 to well over $100,000 per QUALY, suggesting that the widespread drug therapy of elevated cholesterol could not be considered a cost-effective intervention.[105]

In a health care environment increasingly influenced by managed-care formulary decisions, the introduction of cost-effectiveness brackets into national treatment guidelines complicated the process by which pharmaceutical companies promoted cholesterol-lowering medications. Safety and efficacy might satisfy the FDA, but in order to gain access to the broadest possible markets, the pharmaceutical industry recognized that it would need to conduct its own cost-effectiveness research on a much larger scale.[106] Pharmaceutical manufacturers worked to coopt the new field of pharmacoeconomics by incorporating cost-effectiveness outcomes as endpoints for major postmarketing clinical trials. In 1991, Antonio Gotto published a study in the *American Journal of Cardiology* which bounded a small group of "high risk" patients in whom the cost per life-year saved was calculated to be far more cost-effective, at $6,000 to $53,000. This fraction of the total treatable population, which Gotto estimated to number 800,000 in the United States, bore "sufficiently high risk for CAD (coronary artery disease) so that the net cost of lovastatin therapy can be favorably compared with other widely used medical interventions."[107] As cost-effectiveness studies became useful in the marketing of statins, many authors who had originally written critical studies set up a small industry of cost-effectiveness research firms that fielded contract work for the pharmaceutical industry.[108] In the immediate wake of the 4S trial, the *New York Times* reported a Merck-based study claiming that *Zocor* could significantly lower hospital bills when used in secondary prevention. Similar data was gathered during *Pravachol's* subsequent CARE trial and more detailed cost-effectiveness outcomes were incorporated into the primary-prevention trials of *Pravachol* and *Mevacor,* arguing that the drugs were an "economically attractive" remedy averaging $30,000 per life-year saved.[109]

These cost-effectiveness data collected by industry-funded trials were contested by other analysts, who argued that the widespread use of statins was not cost-effective.[110] But even as this critical pharmacoeconomic literature persisted, however, its voice was attenuated by the multiplicity of possible metrics and methodologies of costing life available by the end of the century.[111] In addition to the intricate definitions of the DALY and the QUALY, a range of other pharmacoeconomic indicators came to include the WTP (willingness to pay), the WTGT (willingness to give up leisure time), and the MAR (maximum acceptable risk) among others.[112] Collectively known as "contingent evaluation approaches," these metrics focused on individual preference, rather than systematic public health prioritization, as the base unit of proper economic evaluation, and helped to generate a more industry-friendly perspective on cost-effectiveness. This literature was further supported by surveys of American cardiologists, which suggested a tendency to recommend the use of lipid-lowering therapy, even in cases where it was estimated to cost well over $100,000 per life-year saved.[113] Although cost could suggest a bottom threshold for cholesterol treatment, that threshold would prove to be easily contested and impossible to enforce.[114]

In addition to utilitarian critiques of pharmacopreventive practice, a set of principle-based ethical critiques has also continued to oppose the expanded prescription of statins and loosely work to maintain a lower limit on treatable cholesterol levels. Based on a priori moral principles rather than any standardized calculus of benefit, this family of arguments starts with the premise that routinely medicating a population that is not egregiously sick itself represents a fundamental moral breach that devalues human life and dignity. For the purposes of this discussion, principalist arguments against the wide-scale medication of a population can be divided into arguments of "medicalization" on the one hand and what Mickey Smith has dubbed "pharmacologic Calvinism" on the other.[115]

The medicalization critique is typically a top-down approach which accuses a powerful and interested organization—most frequently the medical profession, the state, or the pharmaceutical industry—of manufacturing a disease and producing populations of patients in order to consolidate control over power and resources.[116] Less frequently, critics of medicalization do not blame a single powerful institution and instead lament more generally the loss of wellness and the broader social costs (beyond risk of adverse events and dollars spent) incurred by defining the human body as inherently diseased instead of inherently healthy.[117] Critics of medicalization have often been able to deftly untangle and expose the links between interest and the definition of disease; however, such analyses rarely trump clinical trials data when decisions are being made at the level of drug regulation, formulary acceptance, or clinical prescription. In the case of the statins, clinicians generally seem to be more concerned with the underutilization of the drugs than with any notion that widespread statin prescription might be culturally or psychologically harmful.

If critiques of medicalization are rooted in political economy, critiques of "pharmacologic Calvinism," focus instead on the morality of the individual consuming the drug.[118] Such arguments reflect suspicions that the decision to seek a pharmaceutical solution as a replacement for some other, more individually responsible solution (e.g., diet, exercise, existential reckoning) reflects a corrosive moral laxity, a short-circuiting between effort and result. In some popular and medical polemics, statins are not seen as agents curing a disease state but rather as technologies of enhancement; physiological crutches which are used to support an immoral lifestyle.[119] The image of the overfed, underexercised American consumer who takes a statin with his cheeseburger has swiftly become something of a cultural cliché.[120] But in a culture of irony, the moral valence of such clichés is easily twisted. Thus, a food critic could, by 2000, offer her highest praise to a restaurant's pâté de fois gras, Chateaubriand, or crème brulée by advising her readers to be sure to take their *Lipitor* before the meal.[121] In this moral inversion, the cure to the latter-day ailments of excess consumption lies, cleverly, not in limiting consumption but in an adjunct prescription. Were it to be approved for the general marketplace, *Mevacor OTC* would represent the ideal extension of a culture of consumption, which finds its ultimate solution to the morbid consequences of overconsumption in the production of another consumable.

In examining the relationship between cholesterol, *Mevacor,* and the other statins, we witness the negotiation of diagnosis at its most abstractly market-oriented: bounded at the bottom by the cost of the product, bounded at the top by a series of large, expensive, and persuasive clinical trials that have influenced guidelines and practice. Ultimately, what shields the remainder of the adult population from being eligible for statin consumption is a very loosely tied network of economic and moral arguments that might yet be altered by another large study or by a shift in regulatory and consumer practices. Though the possibility of adding *Mevacor* to the drinking water still seems quite remote, the possibility of over-the-counter statins has recently been rekindled. By early 2004, newspapers announced that statins would soon be made available over-the-counter in Great Britain, and the American pharmaceutical trade press noted that Merck and Bristol-Myers Squibb were likely to petition the FDA once more to seek over-the-counter status for *Mevacor* and *Pravachol*.[122] After all, it is not impossible in a product's life-cycle for a resurrection to occur. Perhaps, years after its death, *Mevacor* will rise again.

The biography or life-cycle of the pharmaceutical differs from the biography or life-cycle of its human subjects in that whatever agency we might attribute to the pharmaceutical is only understandable in the context of the various intersecting institutions that support and in some ways constitute its existence. This chapter has described the interconnected history of several of these institutions: the approval and indication-granting practices of the Food and Drug Administration, the diagnostic threshold and therapeutic recommendations of the National Cholesterol Education Program, the conduct of clinical trials, and the marketing and clinical research practices of the American pharmaceutical industry. The observation that the goals of the public health advocates of the NCEP easily merged with the goals of the pharmaceutical industry in a shared expansive tendency toward pharmaceutical prevention and a shared marketing of risk to the general population does not suggest that any scandalous influence was exerted on either part. Instead, the sequence of events chronicled in this chapter illustrates precisely how much the pharmaceutical industry has managed to accomplish through the construction of a means of product promotion that seems transparently *licit* at the highest administrative levels. Elevated cholesterol is a disorder of pure number, in which the diagnostic process is now as much a negotiation between the pharmaceutical industry and guideline-setting committees as it is a negotiation between doctor and patient. As this examination of *Mevacor* has demonstrated, the joined career of statins and high cholesterol in the late twentieth century provides a particularly relevant illustration of the fluidity of contemporary boundaries between physiology and pathology.

Notes

1. Ira S. Nash, Lori Mosca, Roger Blumenthal, Michael H. Davidson, Sidney C. Smith, Jr., Richard C. Pasternak, "Contemporary Awareness and Understanding of Cholesterol as a Risk Factor: Results of an American Heart Association National Survey," *Archives of Internal Medicine* 163 (2003)13: 1597–1600; R. M. Pieper, D. K. Arnett, P. G. McGovern, E. Shahar, H. Blackburn, and R. V. Luepker, "Trends in Cholesterol Knowledge and Screening and Hypercholesterolemia Awareness and Treatment, 1980–1992, The Minnesota Heart Study," *Archives of Internal Medicine* 157 (1997)20: 2326–32; Earl S. Ford, Ali H. Mokdad, Wayne H. Giles, George A. Mensah, "Serum Total Cholesterol Concentrations and Awareness, Treatment, and Control of Hypercholesterolemia Among US Adults: Findings From the National Health and Nutrition Survey, 1999 to 2000," *Circulation* 107 (2003): 2185–9.

2. On the role of the clinical trial in the broader history of therapeutic rationalism in the twentieth century, see Harry M. Marks, *The Progress of Experiment: Science and Therapeutic Reform in the United States, 1900–1990* (Cambridge: Cambridge University Press, 1997); a more recent sociological account of standardization and guidelines in American medicine is found in Stefan Timmermans and Marc Berg, *The Gold Standard: The Challenge of Evidence-Based Medicine and Standardization in Health Care* (Philadelphia: Temple University Press, 2003).

3. On psychopharmaceuticals and the definition of psychiatric disorders, see David Healy, *The Antidepressant Era* (Cambridge: Harvard University Press, 1995); and Nathan Greenslit, "Depression and consumption: psychopharmaceuticals, brands, and the new identity politics," *Culture, Medicine, and Psychiatry* (in press); also see Joe Dumit, *Drugs for Life: Managing Identity Through Facts and Pharmaceuticals* (forthcoming). Existing historical works which have looked at the role of pharmaceuticals and the negotiation of somatic disease include Keith Wailoo, *Drawing Blood: Technology and Disease Identity in Twentieth-Century America* (Baltimore: Johns Hopkins University Press, 1997); Chris Feudtner, *Bittersweet: Diabetes, Insulin, and the Transformation of Illness* (Chapel Hill: University of North Carolina Press, 2003); Robert Kaiser, "The Introduction of the Thiazides: A Case-Study in 20th Century Therapeutics," in *The Inside Story of Medicines*, ed. Greg Higby and Elaine Stroud (Madison: American Institute of the History of Pharmacy, 1997), 121–137; Rein Vos, *Drugs Looking for Diseases: Innovative Drug Research and the Development of the Beta Blockers and the Calcium Antagonists* (Dordrecht: Kluwer Academic Publishers, 1990).

4. On the pharmaceutically mediated medicalization of preventive categories, see Maren Klawiter, "Risk, prevention, and the breast cancer continuum: the NCI, the FDA, health activism, and the pharmaceutical industry," *History and Technology* 18(2002): 309–353, on the late twentieth-century history of the "risk factor," see Merwyn Susser, "Epidemiology in the United States after World War II: The

Evolution of Technique," *Epidemiologic Reviews* 7(1985): 147–177, and by Robert Aronowitz, *Making Sense of Illness: Science, Society, and Disease* (New York: Cambridge University Press, 1998), 111–144. For broader perspective on the history of preventive medicine, see George Rosen, *Preventive Medicine in the United States* (New York: Science History Publications, 1975); early publications calling for the identification of asymptomatic disease states include Alexander M. Campbell, "The Necessity for a Periodical Examination of the Apparently Healthy," *Detroit Medical Journal,* 4 (1904): 193–5; Haven Emerson, "The Protection of Health by Periodic Medical Examinations," *Journal of the Michigan State Medical Society,* 21 (1922): 158–171.

5. Arjun Appadurai's framework discussing the different spheres of knowledge-production, knowledge-circulation, and knowledge-consumption surrounding commodities has been quite influential, see Arjun Appadurai, "Introduction: Commodities and the Politics of Value," in Appadurai, ed., *The Social Life of Things: Commodities in Cultural Perspective* (Cambridge: Cambridge University Press, 1988), pp. 3–63.

6. As a series of mergers in the 1980s and 1990s consolidated the industry into a small number of large, publicly traded corporations, pressures mounted for pharmaceutical executives to concentrate their efforts on producing a small number of drugs with large-volume sales, as opposed to a large number of drugs with smaller markets. On the relationship of the sulfa drugs and penicillin to prior moments in the history of the American pharmaceutical industry, see Harry F. Dowling, *Medicines for Man: the Development, Regulation, and Use of Prescription Drugs* (New York: Knopf, 1970); John Lesch, "Chemistry and Biomedicine in an Industrial Setting: The Invention of the Sulfa Drugs," in Seymour H. Mauskopf, ed., *Chemical Sciences in the Modern World* (Philadelphia: University of Pennsylvania Press, 1993), pp. 158–215; and Gladys Hobby, *Penicillin, Meeting the Challenge* (New Haven: Yale University Press, 1985). For broader perspective on the history of the American pharmaceutical industry, see Jonathan Liebenau, *Medical Science and Medical Industry* (Baltimore: Johns Hopkins University Press, 1987).

7. The idea of a drug having a biography or "life-cycle" has been employed in a small number of pharmaceutical histories, e.g., Susan L. Speaker, "From Happiness Pills to National Nightmare: Changing Cultural Assessment of Minor Tranquilizers in America, 1955–1980," *Journal of the History of Medicine* 52 (1997): 338–377; Jordan Goodman and Vivien Walsh, *The Story of Taxol: Nature and Politics in the Pursuit of an Anti-Cancer Drug* (Cambridge: Cambridge University Press, 2001). The concept was also frequently invoked by post-WWII pharmaceutical marketers themselves, e.g., William E. Cox, Jr. "Product Life Cycles as Marketing Models," *The Journal of Business* 40 (1967): 376–81.

8. On Merck's decision to market *Zocor* as a competitor to *Mevacor* in the context of potential patent expiry and growing competition from other brands, see Olivia Williams, Anne-Marie Jacks, Jim Davis, Sabrina Martinez, "Merck(A)

Mevacor," (unpublished case-study, University of Michigan Business School, September 2002); Olivia Williams, Anne-Marie Jacks, Jim Davis, Sabrina Martinez, "Merck(B) Zocor," (unpublished case-study, University of Michigan Business School, September 2002).

9. Rita Rubin, "Drugmakers Prescribe Move: Cholesterol Treatments Could Be First for Chronic Condition to Shift to OTC," *USA Today,* June 27, 2000, 1E.; also see "Merck seeks over-the counter status for anticholesterol drug," *The Boston Globe,* April 7, 1999, E5. Merck had already sought maximal patent extensions through new indications.

10. By the late 1990s, the subject of Rx to OTC switching was gaining increasing attention within the industry as a means of extending brand life. Seventy pharmaceutical agents switched from Rx to OTC between 1975 and 2000, with a joint projected sales of $4.6 billion.

11. *Nonprescription Mevacor™ FDA Advisory Committee Background Information* (July 2000), NDA 21–213, Dockets Office, Food and Drug Administration, Rockville, Maryland.

12. "Drugmakers Prescribe Move," 1D.

13. The psychosomatic literature on the health risks of hypercholesterolemic diagnosis focused on manifest symptoms which patients developed subsequent to diagnosis with hypercholesterolemia—such as dizziness, shortness of breath, and chest pain—with no discernable physiological basis, except for the patients' identification as "sick" following hypercholesterolemic diagnosis. See, for example, Allan S. Brett, "Psychologic effects of the diagnosis and treatment of hypercholesterolemia: Lessons from case studies," *American Journal of Medicine* 91 (1991): 642–7, also see R. C. Lefebvre, K. G. Hursey, R. A. Carleton, "Labeling of participants in high blood pressure screening programs: implications for blood cholesterol screenings," *Archives of Internal Medicine* 148 (1988): 1993–7; T. Tijmstra, "The psychological and social implications of serum cholesterol screening," *International Journal of Risk and Safety in Medicine* 1 (1990): 29–44.

14. See testimony of Jeffrey L. Anderson, *Department of Health and Human Services, Food and Drug Administration, Public Hearing on FDA Regulation of Over-the-Counter Drug Products,* June 29, 2000, FDA Docket 00N-1256, pp.111–12.

15. *Public Hearing on FDA Regulation of Over-the-Counter Drug Products,* June 29, 2000, 7–8.

16. Robert Scott, then head of cardiovascular and metabolic products for Pfizer, Inc., as quoted in Rubin, "Drugmakers Prescribe Move." Data on *Lipitor* market share from IMS Health, "Developments in the Statin Market," March 16, 2000.

17. Testimony of Edward Frohlich, *Public Hearing on FDA Regulation of Over-the-Counter Drug Products,* June 29, 2000, 131–3.

18. Cross-examination, Robert Temple, ibid., 139–140.

19. E.g., B. Waine Kong, CEO of the Association of Black Cardiologists, noted during the hearings that OTC cholesterol drugs would be more accessible to minority patients, a population in whom hypercholesterolemia was relatively undertreated. Victoria Stagg Elliott, "FDA Advisory Committee Vetoes OTC Status for Low-Dose Anti-Cholesterol Drugs," *AMNews,* August 7, 2000. Also see National Council of Negro Women to Jennie C. Butler, Dockets Management Branch, FDA. In FDA Docket OON-1256, Dockets Management Branch (HFA-305), Food and Drug Administration, Rockville, Maryland.

20. A parallel bid by Bristol-Myers Squibb to receive OTC approval for its own statin, *Pravachol,* was rejected the following day. "Panel Retains Status of Cholesterol Drugs," *New York Times,* July 18, 2000, F12; Elizabeth Mechcatie, "FDA Panel Rejects Making Low-Dose Statins Available OTC," *Family Practice News,* September 1, 2000.

21. The FDA Advisory Panel found that Merck demonstrated adequate safety and cholesterol-lowering efficacy of *Mevacor OTC* but failed to demonstrate that it would produce meaningful clinical endpoints in the target consumer population. Food and Drug Administration, *Minutes of the Joint Meeting of Nonprescription Drugs Advisory Committee and the Endocrinological and Metabolic Drugs Advisory Committee,* July 13 2000, NDA 21–213, Dockets Administration, Food and Drug Administration, Rockville, Maryland.

22. The 1980 NAS Report immediately became the subject of an intense public debate over interest and science; several advocates of low-fat, low-cholesterol diets accused the FNB group of being inappropriately influenced by the beef and dairy industry. For a sample of articles in the *New York Times* in the weeks following the report's release, see Lawrence K. Altman, "Report about Cholesterol Draws Agreement and Dissent," *New York Times,* May 28,1980, D18; Jane E. Brody, "Dispute on Americans' Diets," *New York Times,* May 29, 1980; D18; Brody, "When Scientists Disagree, Cholesterol Is in Fat City," *New York Times,* June 1, 1980, E1; "A Confusing Diet of Fact," *New York Times,* June 2, 1980, A18; Karen De Witt, "Scientists Clash on Academy's Cholesterol Advice," *New York Times,* June 20, 1980, A15; Charles E. Rodgers, Jr., "Cholesterol Diets: A New Low," *New York Times,* June 29, 1980, WC18; Victor Herbert, "In Defense of the Cholesterol Report," *New York Times,* July 13, 1980. For subsequent analysis of the FNB debate, see Patricia Hausman, *Jack Sprat's Legacy: The Science and Politics of Fat & Cholesterol* (New York: Richard Marek, 1981); Marion Nestle, *Food Politics: How the Food Industry Influences Nutrition and Health* (Berkeley: University of California Press, 2001).

23. Thomas R. Dawber, Felix Moore, and George V. Mann, "Coronary Heart Disease in the Framingham Study," *American Journal of Public Health* (April, 1957): 3–24; see also Thomas R Dawber, Gilcin F. Meadors, and Felix Moore, "Epidemiologic Approaches to Heart Disease: The Framingham Study," *American Journal of Public Health* 41 (1951): 279–286. For a concise retrospective of the

goals and proceedings of the Framingham study, see Thomas R. Dawber, *The Framingham Study: The Epidemiology of Atherosclerotic Disease* (Cambridge: Harvard University Press, 1980), and Susser, "Epidemiology in the United States after World War II." Other significant epidemiological investigations of cholesterol and heart disease include A. Keys and K. T. Anderson, "The Relationship of the Diet to the Development of Atherosclerosis in Man," in *Symposium on Atherosclerosis* (Washington, DC: National Academy of Sciences, 1954), 181–187; Norman Joliffe, "Fats, Cholesterol, and Coronary Heart Disease: A Review of Recent Progress," *Circulation*, 20 (1959): 109–127; A. Kagan, B. R. Harris, W. Winkelstein, Jr., "Epidemiological Studies of Coronary Heart Disease and Stroke in Japanese Men Living in Japan, Hawaii, and California: Demographic, Physical, Dietary, and Biochemical Characteristics," *Journal of Chronic Disease* 27 (1974): 345–364; W. B. Kannel, W. P. Castelli, T. Gordon, "Cholesterol in the Prediction of Atherosclerotic Disease," *Annals of Internal Medicine* 90 (1979): 85–91; A. Keys, C. Aravanis, H. Blackburn, *Seven Countries: A Multivariate Analysis of Death and Coronary Heart Disease* (Cambridge: Harvard University Press, 1980).

24. On the history of antihypertensive medications in the 1950s, 1960s, and 1970s, and the development of national guidelines (through the National High Blood Pressure Education Program, or NHBPEP) for the treatment of hypertension, see Jeremy A. Greene, "Releasing the Flood Waters: *Diuril* and the Reshaping of Hypertension," *Bulletin of the History of Medicine* (2005) 79(4): 749-94.

25. Lipid Research Clinics Program, "The Lipid Research Clinics Coronary Prevention Trial Results: I. Reduction in Incidence of Coronary Heart Disease," *Journal of the American Medical Association* 251(1984)3: 351–364; Consensus Conference, "Lowering Blood Cholesterol to Prevent Heart Disease," *Journal of the American Medical Association* 253(1985)14: 2080–2086. On the role of this trial in the formation of the National Cholesterol Education Program, see Jeremy A. Greene, "The Therapeutic Transition: Pharmaceuticals and the Marketing of Chronic Disease" (Harvard: PhD diss., 2005); Karin Garrety, "Social Worlds, Actor-Networks and Controversy: The Case of Cholesterol, Dietary Fat, and Heart Disease," *Social Studies of Science* 27 (1997): 727–773; Jonathan R. Cole, "Dietary Cholesterol and Heart Disease: The Construction of a Medical 'Fact,'" in *Surveying Social Life*, ed. Hubert J. O. Gorman (Middletown: Wesleyan University Press, 1988), 437–466.

26. This section analyzes the changing textbook knowledge of cholesterol and atherosclerosis as a reflection of changes in physician attitudes and practices toward its ontological status as disease state and the pragmatic consensus regarding its diagnosis and treatment. As Ludwik Fleck observes, textbook knowledge always trails behind the leading edge of active scientific debate and journal reviews, but as such it is simultaneously (a) more likely to reflect the level of knowledge of the average medical practitioner and (b) a consistent sampling device for measuring changing practice patterns, as textbook revisions

occur on a regular basis, roughly every 3–5 years. The two textbooks used in this analysis, Russell Cecil's *Textbook of Medicine* (now known as the *Cecil-Loeb Textbook of Medicine*) and T. R. Harrison's *Principles of Internal Medicine* (now *Harrison's Principles of Internal Medicine*) took over the canonical role formerly occupied by William Osler's *Principles and Practice of Medicine* around the mid-twentieth century and remain today as two central texts of American general medical practice.

27. *Textbook of Medicine*, 9th ed., ed. Russell L. Cecil, Robert F. Loeb, (Philadelphia, W. B. Saunders, 1955), 702–3; also see Kendall Emerson, Jr., "Xanthomatosis and Lipidosis," *Principals of Internal Medicine*, 3rd ed., ed. T. R. Harrison (New York: McGraw-Hill, 1958), 742–745.

28. Howard A. Eder, "Primary Hyper- and Hypolipidemias," *Cecil-Loeb Textbook of Medicine*, 11th ed., ed. Paul B. Beeson and Walsh McDermott (Philadelphia: W. B. Saunders, 1963), 1334.

29. Ibid., 1333.

30. Ibid., 1332–1335. For a contemporary clinical review of the problems of equating Gaussian conceptions of normality with definition of healthy physiology, see Elias Amador, "Health and Normality," *Journal of the American Medical Association* 232(1975)9: 953–4. "Nowadays," Amador argued, " a value compatible with health is called 'normal,' a term that has led to confusion because of its biological and statistical uses. The two usages, 'normal' indicating health and 'normal' indicating Gaussian, are not synonymous" (quotation p. 954). Amador preferred to distinguish three sorts of normality in medicine: "qualitative normality," "quantitative normality," and "molecular normality." Also see William E. Stempsey, *Disease and Diagnosis: Value-Dependent Realism* (Boston: Kluwer Academic Publishers, 1999).

31. For a discussion of the renal threshold in diabetes, see Claude Bernard, *Lecons sur le diabete et la glycogenese animale* (1877) as translated in Georges Canguilhem, *The Normal and the Pathological*, trans. Carolyn R. Fawcett (New York: Zone Books, 1991[1943]), 70.

32. On the history of statistics, see Loren Graham, *Between Science and Values* (New York: Columbia University Press, 1981); Theodore M. Porter, *The Rise of Statistical Thinking: 1820–1990* (Princeton: Princeton University Press, 1986); Ian Hacking, *The Taming of Chance* (Cambridge: Cambridge University Press, 1990). On the labeling of statistical outliers as medically deviant, see J. Rosser Matthews, *Quantification and the Quest for Medical Certainty* (Princeton: Princeton University Press, 1995); William G. Rothstein, *Public Health and the Risk Factor: A History of an Uneven Medical Revolution* (Rochester: University of Rochester Press, 2003), 36–49.

33. Daniel Steinberg and Antonio Gotto, "Preventing Coronary Artery Disease by Lowering Cholesterol Levels: Fifty Years from Bench to Bedside," *Journal of the American Medical Association* 282 (1999)21: 2043–50.

34. The periodical literature of the '50s and '60s was saturated with material on the epidemic of heart disease and attempts to control it, e.g., "Last ten years: giant steps against heart disease," *Today's Health*, July 1959, 30–33ff.; "Can heart attacks be predicted?" *Today's Health*, December 1960, 36–7ff.; J. Stuart, "Any man can have a heart attack," *Today's Health*, November 1961, 12–13; L. Kavaler, "Will there soon be a drug that might ultimately prolong your husband's life?" *Good Housekeeping*, October 1969, 112–3ff.; "Drug to prevent heart attacks?" *U.S. News & World* Report, July 8, 1968, 13. For a more general exploration of "diseases of civilization" or luxus pathologies in American cultural history, see Allan Brandt, "Behavior, Disease, and Health in the Twentieth-Century United States: The Moral Valence of Individual Risk," and Charles Rosenberg, "Banishing Risk: Continuity and Change in the Moral Management of Disease," in Allan M. Brandt and Paul Rozin, eds., *Morality and Health* (London: Routledge, 1997), 35–52, 53–79; also see Rosenberg, "Pathologies of Progress: The Idea of Civilization as Risk," *Bulletin of the History of Medicine* 72 (1998)4: 714–30.

35. Keys and Anderson, "The Relationship of the Diet to the Development of Atherosclerosis in Man," in *Symposium on Atherosclerosis*; Norman Joliffe, "Fats, Cholesterol, and Coronary Heart Disease"; Kagan, Harris, Winkelstein, Jr., "Epidemiological Studies of Coronary Heart Disease and Stroke in Japanese Men Living in Japan, Hawaii, and California." This is only the most recent chapter in a long history of medical primitivism, in which practices and objects of "pre-industrialized" life are invested with an a priori sense of healthfulness and presented as balms to the harried denizens of the modern age.

36. Consensus Conference, "Lowering Blood Cholesterol to Prevent Heart Disease," quotation p. 2083.

37. Ibid.

38. For example, the 1970 entry in *Harrison's* listed xanthomatosis as a "morphological term" in contrast to the preferred molecular taxonomy of "primary disturbances in lipid metabolism." See Donald S. Fredrickson, "Disorders of Lipid Metabolism and Xanthomatosis," *Harrison's Principles of Internal Medicine*, 6th ed., ed. Maxwell M. Wintrobe, George W. Thorn, Raymond D. Adams, Ivan L. Bennett, Eugene Braunwald, Kurt J. Isselbacher, and Robert G. Petersdorf (New York: McGraw-Hill, 1970), 630.

39. Ibid., 633: "A prompt fall to normal cholesterol levels in a diet restricted in cholesterol and saturated fats and high in polyunsaturated fats is both a good diagnostic test and adequate therapy. It has not yet been proved that hypolipidemic agents are useful or necessary in treatment of acquired type II."

40. "The primary hyperlipoproteinemias can be divided into two broad categories: (1) *single-gene disorders* that are transmitted by simple dominant or recessive mechanisms and (2) *multifactorial disorders* with complex inheritance patterns in which multiple variant genes interact with environmental factors to produce varying degrees of hyperlipoproteinemia in members of a family." Note

that in this system, there is no longer any "acquired" form, only genetics that is well understood (single-gene) and genetics that is murky and poorly understood (multifactorial). Michael S. Brown and Joseph L. Goldstein, "The Hyperlipoproteinemias and Other Disorders of Lipid Metabolism," *Harrison's Principles of Internal Medicine,* 9th ed., ed. Kurt J. Isselbacher, Raymond D. Adams, Eugene Braunwald, Robert G. Petersdorf, and Jean D. Wilson (New York: McGraw-Hill, 1980), 507–518, quotation p. 507.

41. Ibid., 516.

42. Marketing decisions regarding pharmaceutical products in development began long before launch. For example, in early 1985, Jonathan Tobert, head of the *Mevacor* product team inside of Merck, was disturbed to learn that the pharmaceutical staff had created *Mevacor* as a yellow tablet. Protesting that "yellow is the color of butter," and therefore had no place in a cholesterol-reducing product, Tobert mandated that the pharmacologist redesign *Mevacor* as a "soothing" light-blue tablet instead. John A. Byrne, "The Miracle Company," *Business Week,* October 19, 1987, 88.

43. "In its largest effort to date, MSD Marketing faces the challenges and opportunities of launching a major new product and simultaneously developing a previously unsatisfied and undeveloped market." "Cholesterol breakthrough: *Mevacor* caps a decades-long research effort," *Merck World* 8(1987)5: 4–13, quotation p. 13.

44. The 1987 entry in *Harrison's Principles of Internal Medicine* mentions *Mevacor* only in conjunction with FH. "The Hyperlipoproteinemias and Other Disorders of Lipid Metabolism," *Harrison's Principles of Internal Medicine,* 11th ed., ed. Eugene Braunwald, Kurt J. Isselbacher, Robert J. Petersdorf, Jean D. Wilson, Joseph B. Martin, and Anthony S. Fauci (New York: McGraw-Hill, 1987), 1658.

45. "Cholesterol breakthrough: *Mevacor* caps a decades-long research effort," quotation p. 13.

46. The early *Mevacor* clinical trials focused on patients with heterozygous FH, e.g., D. W. Bilheimer, S. M. Grundy, M. S. Brown, J. L. Goldstein, "Mevinolin and colestipol stimulate receptor-mediated clearance of low density lipoprotein from plasma in familial hypercholesterolemia heterozygotes," *Proceedings of the National Academy of Sciences of the United States of America* 80 (1983)13: 4124–8; S. M. Grundy, D. W. Bilheimer, "Inhibition of 3-hydroxy-3-methylglutaryl-CoA reductase by mevinolin in familial hypercholesterolemia heterozygotes: effects on cholesterol balance," *Proceedings of the National Academy of Sciences of the United States of America* 81 (1984): 2538–42; D. R. Illingworth, G. J. Sexton, "Hypercholesterolemic effects of mevinolin in patients with heterozygous familial hypercholesterolemia," *Journal of Clinical Investigation* 74 (1984): 1972–8; J. M. Hoeg, M. B. Maher, L. A. Zech, K. R. Bailey, R. E. Gregg, K. J. Lackner, S. S. Fojo, M. A. Anchors, M. Bojanovski, D. L. Sprecher, et al., "Effectiveness of mevinolin on

plasma lipoprotein concentrations in type II hyperlipoproteinemia," *American Journal of Cardiology* 57 (1986): 933–9; R. J. Havel, D. B. Hunninghake, D. R. Illingworth, R. S. Lees, E. A. Stein, J. A. Tobert, S. R. Bacon, J. A. Bolognese, P. H. Frost, G. E. Lamkin, et al., "Lovastatin (mevinolin) in the treatment of heterozygous familial hypercholesterolemia. A multicenter study," *Annals of Internal Medicine* 107 (1987)5: 609–15. A few case reports by the same investigators also focused on homozygotes, see C. East, S. M. Grundy, D. W. Bilheimer, "Normal cholesterol levels with lovastatin (mevinolin) therapy in a child with homozygous familial hypercholesterolemia following liver transplantation," *Journal of the American Medical Association* 256 (1986): 2843–8;"Cholesterol breakthrough: *Mevacor* caps a decades-long research effort," 12. Most of the studies were conducted through Grundy's Texas group, e.g., C. A. East, S. M. Grundy, D. W. Bilheimer, "Preliminary report. Treatment of type 3 hyperlipoproteinemia with mevinolin," *Metabolism* 35 (1986): 7–8; G. L. Vega, C. A. East, S. M. Grundy, "Lovastatin therapy in familial dysbetalipoproteinemia: effect on kinetics of apoprotein B," *Atherosclerosis* 70 (1988): 131–143; A. Garg and S. M Grundy, "Lovastatin for lowering cholesterol levels in noninsulin-dependent diabetes mellitus," *New England Journal of Medicine* 314 (1988): 81–6; G. L. Vega and S. M. Grundy, "Lovastatin therapy in nephritic hyperlipidemia: effects on lipoprotein metabolism," *Kidney* 33 (1988): 1160–8.

47. "Cholesterol breakthrough: *Mevacor* caps a decades-long research effort," 12.

48. Alfred Alberts, oral history conducted by Edward Shorter, tape 1 side 1 of "Lovastatin lunch," February 1987; from "The Health Century" oral history collections NLM, OH 136, History of Medicine Division, National Library of Medicine, Bethesda, Maryland.

49. Merck's investment in non-FH clinical trials began only shortly after favorable results began to appear supporting *Mevacor* in FH, through the efforts of a group called "The Lovastatin Study Group." See Lovastatin Study Group II, "Therapeutic response to lovastatin (mevinolin) in nonfamilial hypercholesterolemia. A multicenter study," *Journal of the American Medical Association* 256 (1986)20: 2829–34; Lovastatin Study Group III, "A multicenter comparison of lovastatin and cholestyramine therapy for severe primary hypercholesterolemia," *Journal of the American Medical Association* 260 (1988)3: 359–66.

50. E.g., Philip J. Hilts, "FDA Approves Sale of More Effective Cholesterol-Lowering Drug," *The Washington Post,* September 2, 1987, A3.

51. Gotto's extensive institutional affiliations are discussed in Timothy J. Moore, "The Cholesterol Myth," 54.

52. Hilts, "FDA Approves Sale of More Effective Cholesterol-Lowering Drug."

53. Mike Gorman to Antonio Gotto, memorandum, c.1985, found in box 2, folder 3, Mike Gorman Papers, MS C 462, National Library of Medicine, Bethesda, Maryland (hereafter referred to as "Gorman Papers").

54. "Cholesterol breakthrough: *Mevacor* caps a decades-long research effort," 13.

55. For study design, see Reagan H. Bradford, Charles L. Shear, Athanassios N. Chremos, Carlos Dujovne, Frank A. Franklin, Michael Hesney, Jim Higgins, Alexandra Langendorfer, James L. Pool, Harold Schnaper, Wendy P. Stephenson, "Expanded clinical evaluation of lovastatin (EXCEL) study: design and patient characteristics of a double-blind, placebo controlled study in patients with moderate hypercholesterolemia," *American Journal of Cardiology* 66 (1990): 44B–55B.

56. Additional goals for the EXCEL trial included extended safety monitoring for cataracts and myopathy and comparison of the effectiveness of variable dose-regimens.

57. Michael F. Oliver, "Consensus or Nonsensus Conferences on Coronary Heart Disease," *Lancet* (1985): 1087–9; Edward H. Ahrens, "The Diet-Heart Question in 1985: Has It Really Been Settled?" *Lancet* (1985): 1085–7.

58. R. H. Bradford, C. L. Shear, C. Dujovne, M. Downton, F. A. Franklin, A. L. Gould, M. Hesney, J. Higgins, D. P. Hurley, et al., "Expanded clinical evaluation of lovastatin (EXCEL) study results. I. Efficacy in modifying plasma lipoproteins and adverse event profile in 8245 patients with moderate hypercholesterolemia," *Archives of Internal Medicine* 151 (1991)1: 43–9; R. H. Bradford, C. L. Shear, A. N. Chremos, F. A. Franklin, D. T. Nash, D. P. Hurley, C. A. Dujovne, J. L. K. Pool, H. Schnaper, H. Hesney, et al., "Expanded clinical evaluation of lovastatin (EXCEL) study results: III. Efficacy in modifying lipoproteins and implications for managing patients with moderate hypercholesterolemia," *American Journal of Medicine* 91(1991)1B: 18S–24S.

59. Furthermore, even though the NCEP guidelines only called for the use of statins in the 240/130 population—the rest being advised to employ dietary and behavioral change—the gap between the pathological level of 240 and the ideal level of 200, coupled with the fact that most practicable diets, even when effective, typically only lowered blood cholesterol by 10% or less, helped to guarantee a wider role for *Mevacor* than might be apparent from a superficial reading of the NCEP flowchart. This argument had been made in support of broader *Mevacor* usage by product manager Jonathan Tobert as early as 1987; see J. A. Tobert, "New developments in lipid-lowering therapy: the role of inhibitors of hydroxymethylglutaryl-coenzyme A reductase," *Circulation* 76 (1987)3: 534–538.

60. National Cholesterol Education Program, *Second Report of the Expert Panel on Detection, Evaluation, and Treatment of High Blood Cholesterol in Adults (Adult Treatment Panel II)* (Bethesda, MD: National Institutes of Health, National Heart, Lung, and Blood Institute, 1993), III 3–III10.

61. "Bristol-Myers Gets Approval on Drug," *New York Times,* November 2, 1991, 39.

62. R. H. Brandford et. al., "Expanded clinical evaluation of lovastatin (EXCEL) study: design and patient characteristics," 45B.

63. Adriana Petryna, "The Human Subjects Research Industry," in Adriana Petryna, Andrew Lakoff, and Arthur Kleinman (eds.) *Global Pharmaceuticals: Ethics, Markets, Practices* (Durham, NC: Duke University Press, 2006). For comments on the early history of the CRO industry, see J. E. Beach, "Clinical Trials Integrity: A CRO Perspective," *Accountability in Research* 8 (2001)33: 245–60; also see R. K. H. Wyse and R. G. Hughes, *Pharmaceutical Contract Research in the 1990s* (London: Technomark Consulting Services, 1993). The emergence of a more comprehensive client-centered approach is evident in David Fin, "Temporary Help Stretches to Long Term: Contract Research Organizations, Outsourcing Lets Consolidating Companies Cut Costs and Eliminate Excess Capacity," *Financial Times*, July 15, 1999. The slogan of Quintiles Transnational, the world's largest CRO during the mid-1990s, was "Follow the Molecule," emphasizing the firm's full capacity for providing services to guide a client's drug through any stage of the development process.

64. Fin, "Temporary Help Stretches to Long Term," 5. Only five years earlier the number of CROs was only 1,000 worldwide, and the total market size at $3 billion. Clive Cookson, "Survey of Pharmaceuticals," *Financial Times*, March 23, 1994.

65. K. A. Getz, *AMCs Rekindling Clinical Research Partnerships with Industry* (Boston: Centerwatch, 1999), as cited in Thomas Bodenheimer, "Uneasy Alliance: Clinical Investigators and the Pharmaceutical Industry," *New England Journal of Medicine* 342 (2000)20: 1539–1544, p. 1540.

66. Frank Hurley, as quoted in Kathleen Day, "Test-Driving Pharmaceuticals: Biometric Research Finds Success in Saving Drug Manufacturers Time and Money," *The Washington Post*, March 21, 1994, F5; also see Cookson, "Survey of Pharmaceuticals."

67. E.g., M. R. Weir, M. I. Berger, C. L. Liss, N. C. Santanello, "Comparison of the effects on quality of life and of the efficacy and tolerability of lovastatin versus pravastatin: The quality of life multicenter group," *American Journal of Cardiology* 71 (1993): 810–815. Merck also conducted an analogous fleet of studies comparing its new product *Zocor* to *Pravachol* with equally favorable results for Merck. See "Merck claims efficacy advantage for statins," *Pharma Marketletter*, June 1, 1992.

68. Analysis of competitive trials data from Cathy Kelley, Mark Helfand, Chester Good, and Michael Ganz, *Drug Class Review: Hydroxymethylglutaryl-coenzyme-A Reductase Inhibitors (statins)* (Washington, DC: Veterans Health Administration, Pharmacy Benefit Management, Strategic Healthcare Group), 3–10.

69. National Cholesterol Education Program, *Second Report of the Expert Panel on Detection, Evaluation, and Treatment of High Blood Cholesterol in Adults (Adult Treatment Panel II): Executive Summary*, 14–20.

70. Both Merck and BMS initially began their secondary prevention trials in the late 1980s, well before the NCEP guidelines were revised.

71. *Zocor* recorded sales of $700 million in 1992, a year in which *Mevacor* was already capturing over $1 billion in revenue. Williams et al., "Merck(B) *Zocor*," 2.

72. For the design of the 4S trial, see Scandinavian Simvastatin Survival Study Group, "Design and baseline results of the Scandinavian Simvastatin Survival Study of patients with stable angina and/or previous myocardial infarction," *American Journal of Cardiology* 71 (1993)5: 393–400; Scandinavian Simvastatin Survival Study Group, "Randomized trial of cholesterol lowering in 4444 patients with coronary heart disease: the Scandinavian Simvastatin Survival Study (4S)," *Lancet* 344 (1994)8934: 1383–89. For a review of the study's significance see S. Guptha, "Profiling a landmark clinical trial: Scandinavian Simvastatin Survival Study," *Current Opinion in Lipidology* 6(5) 1995: 251–3.

73. Clive Cookson, "Drugs Improve Heart Patient's Prospects," *Financial Times*, November 18, 1994, p.4; Gina Kolata "Cholesterol Drugs Found to Save Lives," *New York Times*, November 17, 1994, A1.

74. "FDA Approves *Zocor* Label Change," *Pharma Marketletter*, July 10, 1995; Williams et al., "Merck(B) *Zocor*."

75. Estimated cost of trial from "Wider Use Seen for Treatment of Cholesterol," *New York Times*, March 27, 1996, A17. For the CARE trial, see Frank M. Sacks, Marc A. Pfeffer, Lemuel A. Moye, Jean L. Rouleau, John D. Rutherford, Thomas G. Cole, Lisa Brown, J. Wayne Warnica, J. Malcolm O. Arnold, Chuan-Chuan Wun, Barry R. Davis, and Eugene Braunwald, "The Effects of Pravastatin on Coronary Events After Myocardial Infarction in Patients with Average Cholesterol Levels," *New England Journal of Medicine* 335 (1996)14: 1001–9.

76. National Cholesterol Education Program, *Cholesterol Lowering in the Patient with Coronary Heart Disease* (Bethesda: NHLBI, 1997), 4.

77. "Statin Trial Results Imminent," *Pharmaceutical Business News*, September 27, 1995.

78. Ron Winslow and Elyse Tanouye, "Bristol-Myers Enters Battle of Heart Drugs," *Wall Street Journal*, November 16, 1995, B1; J. Shepherd, S. M. Cobbe, I. Ford, et al., "Prevention of coronary heart disease with pravastatin in men with hypercholesterolemia," *New England Journal of Medicine* 333 (1995)20: 1301–7.

79. Jane E. Brody, "Benefit to Healthy Men Is Seen From Cholesterol-Cutting Drug: Study Finds Reduced Incidence of Heart Attack," *New York Times*, November 16, 1995, A1.

80. Torje Pedersen, "Lowering Cholesterol with Drugs and Diet," *New England Journal of Medicine* 333 (1995): 1350–1.

81. "F.D.A. Allows Drug as a Heart Medicine," *New York Times*, July 9, 1996; "Bristol-Myers Squibb Co.: FDA approves ad claims on drug, first heart attack," *Wall Street Journal*, July 9, 1996, B6.

82. "*Pravachol* Helps Prevent First Heart Attacks," two-page advertisement, *The New York Times*, first printed September 8, 1996, 33–34. Promotion of *Mevacor* and *Pravachol* utilized direct-to-consumer (DTC) media a decade before the 1997

decision, see Milton Liebman, "DTC's Role in the Statin Bonanza," *Medical Marketing and Media* 36 (2001)11: 86–90.

83. Thomas A. Pearson, "Lipid-Lowering Therapy in Low-Risk Patients," *Journal of the American Medical Association* 279 (1998)20: 1659–61.

84. J. R. Downs, P. A. Beere, E. Whitney, M. Clearfield, S. Weis, J. Rochen, E. A. Stein, D. R. Shapiro, A. Langendorfer, A. M. Gotto, Jr., "Design and rationale of the Air Force/Texas Coronary Atherosclerosis Prevention Study (AFCAPS/TexCAPS)," *American Journal of Cardiology* 80 (1997)3: 287–93; J. R. Downs, M. Clearfield, S. Weis, E. Whitney, D. R. Shapiro, P. A. Beere, A. Langendorfer, E. A. Stein, W. Kruyer, A. M. Gotto, Jr., "Primary prevention of acute coronary events with lovastatin in men and women with average cholesterol levels: results of AFCAPS/TexCAPS. Air Force/Texas Coronary Atherosclerosis Prevention Study," *Journal of the American Medical Association* 279 (1998): 1615–22.

85. "*Mevacor* Study May Extend Lipid-Lowering Therapy," *Pharma Marketletter*, November 13, 1997.

86. "*Mevacor* Approved for Primary Prevention," *Pharma Marketletter*, March 18, 1999.

87. Margot Mellies and Hartmann Willhoefer, "Current and future clinical trials," in Allan Gaw, Christopher J. Packard, James Shepherd (eds.) *Statins: The HMG CoA Reductase Inhibitors in Perspective* (Malden, MA: Martin Dunitz, 2000), 179–201.

88. National Cholesterol Education Program, *Third Report of the National Cholesterol Education Program (NCEP) Expert Panel on Detection, Evaluation, and Treatment of High Blood Cholesterol in Adults (Adult Treatment Panel III)* (Bethesda: National Institutes of Health, NHLBI, 2001).

89. Thomas M. Burton and Chris Adams, "New U.S. Guidelines Would Triple Use of Cholesterol-Lowering Drugs," *Wall Street Journal*, May 16, 2001, B1.

90. Testimony of Jeffrey L. Anderson, Merck Research Laboratories, *Public Hearing on FDA Regulation of Over-the-Counter Drug Products*, June 29, 2000, 108–11.

91. Rita Rubin, "Drugmakers Prescribe Move: Cholesterol Treatments Could Be First for Chronic Condition to Shift to OTC," *USA Today*, June 27, 2000; Victoria Stagg, "FDA advisory committee vetoes OTC status for low-dose anti-cholesterol drugs," *AMNews*, August 7, 2000.

92. Testimony of Jerome D. Cohen, St. Louis University, *Public Hearing on FDA Regulation of Over-the-Counter Drug Products*, June 29, 2000, 84–5.

93. Testimony of Jeffrey L. Anderson, Merck Research Laboratories, *Public Hearing on FDA Regulation of Over-the-Counter Drug Products*, June 29, 2000, 107.

94. "Serum total cholesterol (mg/dL) levels for persons 20 years of age or older, United States, 1988–94," in *Third Report of the National Cholesterol Education Program (NCEP) Expert Panel on Detection, Evaluation, and Treatment of High Blood Cholesterol in Adults*, III-A-1.

95. E.g., the NHLBI held a conference in 1990 on "Low Blood Cholesterol: Mortality Associations," see Stephen B. Hulley, Judith M. B. Walsh, Thomas B. Newman, "Health Policy on Blood Cholesterol: Time to Change Directions," *Circulation* (1992): 1026–9; also George Davey Smith, Juha Pekkanen, "Should there be a moratorium on the use of cholesterol lowering drugs?" *British Medical Journal* 304 (1992): 431–4. Arguments on the risk of low cholesterol and possible harm via cholesterol lowering were directly countered by proponents of the major statin prevention trials, e.g., Peter H. Jones, "Low Serum Cholesterol Increases the Risk of Noncardiovascular Events: An Antagonist Viewpoint," *Cardiovascular Drugs and Therapy*, 8 (1994): 871–4. By the end of the 1990s, "hypocholesterolemia" contained only the briefest entry in medical textbooks, and the point at which cholesterol levels were low enough to be a possible threat to the organism were defined at under 100mg/dL, a number so far below the range of the general population as to offer no practical lower boundary to the diagnosis of "treatable" cholesterol levels.

96. Scott Allen and Stephen Smith, "Statin Nation: Do we all need to lower our 'bad' cholesterol?" *Boston Globe,* March 16, 2004; Ron Winslow, "Blood Feud: For Bristol Myers, Challenging Pfizer Was Big Mistake; In Rare Head-to-Head Study, *Lipitor* Beats *Pravachol* at Reducing Heart Risk; a New Push on Cholesterol," *Wall Street Journal,* March 9, 2004, A1.

97. See Michael F. Oliver, "Reducing Cholesterol Does Not Reduce Mortality," *Journal of the American College of Cardiologists* 12 (1988)3: 14–7; Michael F. Oliver, "Might treatment of hypercholesterolaemia increase non-cardiac mortality?" *Lancet* 337 (1991): 1529–31; Oliver, "Doubts about preventing coronary heart disease," *British Medical Journal* 304 (1992): 393–4; cf. Uffe Ravbskov, *The Cholesterol Myths: Exposing the Fallacy That Saturated Fat and Cholesterol Cause Heart Disease* (Washington, DC: New Trends Publishing, 2000).

98. Statins do bear a known risk of liver toxicity and rhabdomyolysis, but this has been readily incorporated into existing treatment guidelines with use of diagnostic tests to screen and monitor patients' liver enzymes for any adverse reaction; as such they do not significantly curtail the potential market for the drugs.

99. The Toronto Working Group on Cholesterol Policy, "Asymptomatic Hypercholesterolemia: A Clinical Policy Review," *Journal of Clinical Epidemiology* 43 (1990)10: 1021–1122. See Alan M. Garber, "Where to draw the line against cholesterol," *Annals of Internal Medicine* 111 (1989): 625–27; also see A. M. Garber, B. Littenberg, H. C. Sox, M. E. Gluck, J. L. Wagner, and B. M. Duffy, *Costs and Effectiveness of Cholesterol Screening in the Elderly,* U.S. Congress, Office of Technology Assessment (Washington, DC.: U.S. Government Printing Office, 1989).

100. Toronto Working Group on Cholesterol Policy, "Asymptomatic Hypercholesterolemia," 1093.

101. A Williams, "Screening for risk of CHD: Is it a wise use of resources?" in *Screening for Risk of Coronary Heart Disease,* ed. M. Oliver, M. Ashley Miller, D.

Wood (London: John Wiley & Sons, 1986), 97–106; G. Oster, A. M. Epstein, "Cost-effectiveness of antihyperlipidemic therapy in the prevention of coronary heart disease: the case of cholestyramine," *Journal of the American Medical Association* 258(1987): 2381–7. Also see William Taylor, "Screening for High Blood Cholesterol and Other Lipid Abnormalities," in *Guide to Clinical Preventive Services: Report of the U S. Preventive Services Task Force,* 2nd ed. (Baltimore: Williams & Wilkins, 1989). Though the general literature on cost-effectiveness seems to begin in the late 1970s, e.g., M. C. Weinstein, W. B. Stason, "Foundations of cost-effectiveness analysis for health and medical practices," its use by the mid-1980s was still limited to a relatively small section of the medical literature, e.g., "P. Doubilet, M. C. Weinstein, and B. J. McNeil, "Use and misuse of the term 'cost-effective' in medicine," *New England Journal of Medicine* 314 (1986): 253–6; G. W. Torrance, "Measurement of health states utilities for economic appraisal: review," *Journal of Health Economics* 5 (1986): 1–30; D. U. Himmelstein, S. Woolhandler, "Free care, cholestyramine, and health policy," *New England Journal of Medicine* 311(1984): 1511–14; M. C. Weinstein, W. B. Stason, "Cost-effectiveness of interventions to prevent or treat coronary heart disease," *Annual Review of Public Health* 6 (1985): 41–63.

102. Toronto Working Group on Cholesterol Policy, "Asymptomatic Hypercholesterolemia," 1100. Other critical cost-benefit analyses from this period include W. B. Stason, "Costs and benefits of risk factor reduction for coronary heart disease: insights from screening and treatment of serum cholesterol," *American Heart Journal* 119 (1990): 718–24; M. D. Kelley, "Hypercholesterolemia: the cost of treatment in perspective," *Southern Medical Journal* 83 (1990)12: 1421–5; L. Goldman, M. C. Weinstein, P. A. Goldman, L. W. Williams, "Cost-effectiveness of HMG-CoA reductase inhibition for primary and secondary prevention of coronary heart disease," *Journal of the American Medical Association* 265 (1991): 1145–51; Jonathan S. Silberberg and David A. Henry, "The benefits of reducing cholesterol levels: the need to distinguish primary from secondary prevention," *Medical Journal of Australia* 155 (1991): 665–674; Ivar S. Kristiansen, Anne E. Eggen, Dag S. Thelle, "Cost effectiveness of incremental programmes for lowering serum cholesterol concentration: is individual intervention worth while?" *British Medical Journal* 302 (1991): 1119–22.

103. *Second Report of the Expert Panel on Detection, Evaluation, and Treatment of High Blood Cholesterol in Adults (Adult Treatment Panel II): Final Report,* IV-13.

104. Ibid., IV-15; for the origins of these thresholds in the field of health economics, see M. C. Weinstein, "From cost-effectiveness ratios to resource allocation: where to draw the line?" in *Valuing Health Care: Costs, Benefits, and Effectiveness of Pharmaceuticals and Other Medical Technologies,* ed. F. A. Sloan (New York: Cambridge University Press, 1995).

105. The NCEP's evaluation on this level had been published a year earlier as L. Goldman, D. J. Gordon, B. M. Rifkind, S. B. Hulley, A. S. Detsky, D. W. Good-

man, B. Kinosian, M. C. Weinstein, "Cost and health implications of cholesterol lowering," *Circulation* 85 (1992)5: 1960–8. Note that the authors represent a combination of NCEP advocates (Rifkind, Goodman) and those whose literature had been essential in bringing the cost-effectiveness critique to bear upon the NCEP (Goldman, Kinosian, Weinstein).

106. J. Lyle Bootman, Raymond J. Townsend, William F. McGhan, *Principles of Pharmacoeconomics* (Cincinnati: Harvey Whitney Books, 2001).

107. J. W. Hay, E. H. Wittels, A. M. Gotto, Jr., "An economic evaluation of lovastatin for cholesterol lowering and coronary artery disease reduction," *American Journal of Cardiology* 67 (1991)9: 789–96, quotation p. 789.

108. E.g., D. Thompson and G. Oster, "Cost-effectiveness of drug therapy for hypercholesterolaemia: a review of the literature," *Pharmacoeconomics* 2 (1992)1: 34–42; the authors were now attached to a Brookline, Massachusetts-based company called Policy Analysis Inc., and this very favorable review emphasized the relative cost-effectiveness of *Mevacor* and *Zocor* over cholestyramine.

109. "Cholesterol Pill Linked to Lower Hospital Bills," *New York Times*, March 27, 1995; for examples of cost-effectiveness research incorporated into secondary and primary statin prevention trials, see R. B. Goldberg, M. J. Mellies, F. M. Sacks, et al., "Cardiovascular events and their reduction with pravastatin in diabetic and glucose-intolerant myocardial infarction survivors with average cholesterol levels: subgroup analyses in the Cholesterol and Recurrent Events (CARE) Trial," *Circulation* 98 (1998): 2513–19; J. Caro, W. Klittich, A. McGuire, et al., "The West of Scotland Coronary Prevention Study: weighing the costs and benefits of primary prevention with pravastatin," *British Medical Journal* 315 (1997): 1577–84; A. M. Gotto, S. J. Boccuzzi, J. R. Cook, C. M. Alexander, J. B. Roehm, G. S. Meyer, M. Clearfield, S. Weis, E. Whitney, "Effect of lovastatin on cardiovascular resource utilization and costs in the Air Force/Texas Coronary Atherosclerosis Prevention Study (AFCAPS/TexCAPS)" *American Journal of Cardiology* 86(2000)11: 1176–81.

110. E.g., David Atkins and Carolyn DiGuiseppi, "Screening for High Blood Cholesterol and Other Lipid Abnormalities," in *Guide to Clinical Preventive Services: Report of the U.S. Preventive Services Task Force*, 2nd ed., ed. Carolyn DiGuiseppi, David Atkins, and Steven H. Woolf (Baltimore: Williams & Wilkins, 1996), 15–38; L. A. Prosser, A. A. Stinnett, P. A. Goldman, et al., "Cost-effectiveness of cholesterol-lowering therapies according to selected patient characteristics," *Annals of Internal Medicine* 132(2000): 769–79. The latter study was funded by a grant from the Agency for Healthcare Research and Quality.

111. Ibid.

112. E.g., M. Johannesson, "Economic evaluation of lipid lowering—a feasibility test of the contingent evaluation approach," *Health Policy* 20(1992)3: 309–20.

113. J. M. Gaspoz, J. W. Kennedy, E. J. Orav, L. Goldman, "Cost-effectiveness of prescription recommendations for cholesterol-lowering drugs: a survey of a rep-

resentative sample of American cardiologists," *Journal of the American College of Cardiology* 27 (1996)5: 1232–7.

114. M. Mitka, "Expanding statin use to help more at-risk patients is causing financial heartburn," *Journal of the American Medical Association* 290(2003)17: 2243–5.

115. Mickey Smith, *Small Comfort: A History of the Minor Tranquilizers* (New York: Praeger, 1985).

116. For a review of medicalization theories and risk, see Klawiter, "Risk, prevention, and the breast cancer continuum: the NCI, the FDA, health activism, and the pharmaceutical industry."

117. E.g., Clifton Meador, "The last well person," *New England Journal of Medicine* 330(1994): 440–1; Victoria Stagg Elliott, "Are we all sick? Doctors debate the 'medicalization' of life," *AMNews*, September 20, 2004. On the broader shift from views of the body as inherently healthy to inherently ill, see Joseph Dumit's forthcoming book, *Drugs for Life: Managing Identity with Facts and Pharmaceuticals.*

118. The concept of "pharmacological Calvinism," initially coined by Gerald Klerman, was subsequently popularized by Mickey C. Smith's *Small Comfort: A History of the Minor Tranquilizers* (New York: Praeger Scientific, 1985).

119. For historical and philosophical accounts of the enhancement problem, see David Rothman and Sheila Rothman, *The Pursuit of Perfection: The Promise and Perils of Medical Enhancement* (New York: Random House, 2004); Carl Elliott, *Better than Well: American Medicine Meets the American Dream* (New York: W. W. Norton, 2003).

120. See, e.g., David Noonan, "You want statins with that?" *Newsweek*, July 14, 2004. Similar concerns that pharmaceutical developments might lead to a laxity of regimen can be found in the history of diabetes; see, for example, Chris Feudtner, "The Want of Control: Ideas, Innovations, and Ideals in the Modern Management of Diabetes Mellitus," *Bulletin of the History of Medicine* 69(1995): 66–90.

121. E.g., Sylvia Carter, "New York Strip House," *New York Newsday,* November 23, 2001; John S. Long "Friday!" *Cleveland Plain Dealer,* October 1, 2004, 20.

122. Al Branch, Jr., "J&J/Merck Seek OTC Status for *Mevacor*," *Pharmaceutical Executive,* November 2002, 22.

8

V I A G R A

Making Viagra
From Impotence to Erectile Dysfunction

Jennifer R. Fishman

In the first five years of its availability on the market, Viagra netted approximately $7.4 billion in total sales for Pfizer Incorporated. It is estimated that over 20 million men worldwide have received prescriptions for the drug. Accurately labeled a blockbuster drug, Viagra was colossally successful from the moment of its release; in its first three months of availability in the United States, $411 million worth of Viagra was sold, and 2.7 million prescriptions were written for the drug. The popular press represented Viagra's development as a spontaneous, welcome, and unexpected scientific discovery. In fact, the folkloric tale often told is that sildenafil's (Viagra's chemical name) abilities for enhancing erections were indeed an unintended side effect of the drug, uncovered during a clinical trial investigating its potential for treating angina pectoris (a heart condition which can lead to coronary angina). It was thereafter, when the angina trials failed, that Pfizer pursued bringing the drug to market to treat "erectile dysfunction."

This story, however, is only a partial view of the development of Viagra. While it may be true that Pfizer was not looking to develop an impotence drug, urologists had been looking for medical treatments for decades through a research agenda dedicated to understanding erections (and the absence of them) as a physiological phenomenon. This chapter examines the emergence of Viagra as a treatment for erectile dysfunction within the context of the medicalization of impotence over the last half of the twentieth century, including the categorical transformation of the male sexual disorder "impotence" into "erectile dysfunction" (ED) and from a psychological condition to a medical problem. Through this shift, impotence was not only medicalized, but it was also transformed into an exemplary "lifestyle" condition—a key example of how medical discourses and treatments can now legitimately be used to treat a life-limiting (as opposed to life-threatening) condition. Historically, ideas about impotence and then erectile dysfunction accumulated over time, creating an environment ripe for Viagra's widespread acceptance as a credible treatment for men's sexual problems. Concomitant with the medicalization and "organicization" of erectile (dys)function has been a continued molecularization of the body, in which men's erections and sexual arousal became refined, reduced, and delimited to the functioning of neurotransmitter pathways. "Organicization" is used purposely to refer to the fact that not only is impotence now seen as a condition with *organic* bases, but also that the scope of the disorder itself becomes limited to the penile *organ*.

Locating Viagra's development in the decades prior to its commercial release is important for a number of reasons. It demystifies the rhetoric of Viagra as a "wonder drug," enacting overnight transformations for male sexuality. This analysis seeks to rupture mystifying notions of the "spontaneous" and "novel" implicit in much of the discourse of biomedical innovation.[1] Viagra developed within the context of long-standing efforts by medical professionals and others to place male sexuality under the medical gaze. For this reason, this chapter asks not how Viagra as an emergent technology has transformed our understanding of impotence, but rather, the inverted question: What is the historical context that enabled Viagra to be considered a viable and legitimate treatment for men's sexual problems? Viagra is a culminating event in a longer trajectory of scientific work on the biochemistry of erections. In fact, acceptance of Viagra as a way to treat impotence depended in part upon a shift in knowledge claims about the physiological etiology of the "problem" of erectile dysfunction and the broader acceptance of this biomedical shift. Furthermore, Viagra's success was reliant on networks, discourses, and organizations that emerged in the decades preceding its development. This historicized understanding of Viagra's emergence reconfigures it as part of a nexus of biomedicalizing developments within the production of scientific knowledge about male sexuality.[2]

The Psychologization of Impotence

In the early to mid-twentieth century, impotence was explained primarily as a biological event that accompanied aging in men.[3] Unlike contemporary theories of erectile dysfunction which uncouple impotence from the "normal" aging process, in the early 1900s, decline in sexual function for men was considered part and parcel of the normal aging process. Men had a choice of either accepting and adjusting to these changes or consuming morally and medically questionable elixirs or nostrums such as Brown-Séquard's "Pohl's Spermine Preparations."[4] With the emergence of gerontology (and sexology) as fields of study in the early twentieth century, discourses about "positive aging" began to emerge, touting scientific developments as the path to successful aging, through ideals of vitality, activity, autonomy, and well-being. These ideals still persist in contemporary understandings of erectile dysfunction. However, discourses about "positive aging" were stymied in part due to claims in gerontology and sexology that made sexual decline an element of middle-age. In order to develop a "legitimately positive discourse on sexual decline," impotence had to be transformed from an organic disorder into a psychological one.[5]

Psychological explanations for impotence were considered as early as the 1920s. For example, Sigmund Freud's colleague, psychoanalyst William Stekel, published *Impotence in the Male* in 1927.[6] By midcentury, psychological causes of impotence pervaded the literature. However, this understanding was restricted to *younger* men with impotence. As Marshall and Katz summarize, "impotence was still largely regarded as a disease of the young, and a condition of the old."[7] Therefore, psychotherapy was recommended (and considered successful) for men under 45, while impotence in men over 45 was still considered a "normal" part of the physical decline of aging.

The psychological paradigm, however, quickly spread to include aging men as well, with explanations for impotence ranging from "anxiety" over their supposed loss of sexual function to their "fear" of loss of potency. From the 1960s through the 1980s, ideas about positive aging and positive sexuality took hold, and men were often told that psychological factors were primarily responsible for loss of sexual function and that to *cease* having sex would hasten aging itself. Sexological reports emerging in the mid-twentieth century contributed to this conception of both the importance of one's continued sexual activity over the life course and the normalization of sexuality itself. The Kinsey Report and Masters and Johnson's books normalized sexuality by demystifying it for an American audience.[8] Through discourses about the "naturalness" of sexual activity, Masters and Johnson further adhered to a notion of a universal sexuality, which implies not only a homogeneity of men's and women's experiences, but also an imperative to be sexual.[9] These techniques are "normalizing," not in the sense that they produce bodies or objects that conform to a particular type, but rather that they create a standard model against which objects and actions are judged.[10] Masters and Johnson's work (and I would argue the work of sexology in general)[11] along with others including the *Hite Report* each mentioned that sexual activity was enjoyed throughout the life course.[12] Intended to conceptualize and legitimate older people as sexual beings, they also contributed to what amounts to a mandate for sexual activity for everyone. If one does not measure up against these norms, one may feel inadequate and abnormal. When aligned with the "successful aging" movement in gerontology, sexual activity became a necessary and healthy component of the "good life" as one gets older, though it may be curtailed through psychological blocks stemming from fear and anxiety.

Despite the physiological emphasis of the Masters and Johnson human sexual response cycle, William Masters himself thought that most impotence cases have psychogenic etiologies, as indicated by the modes of behavioral therapy at his institute. This thinking became the widely accepted view of impotence. During the 1970s, a statistic began to circulate in much of the psychological and popular literature on impotence: "At least 90 percent of impotence" is psychologically based. However, the exact origins of this statistic are unclear and it appears not to have been generated by any empirical study.[13] Untethered from whatever roots it had in "empirical research," this statistic traveled across a number of sources, from psychology and sex therapy books and journals and medical articles to self-help books and mainstream magazine articles.[14] Because it decoupled impotence from the "normal aging process," establishing impotence as a psychological condition proved to be an important step in paving the way for Viagra's eventual acceptance as a legitimate treatment for impotence. Impotence was no longer considered a component of aging, but rather an avoidable and even preventable condition created by psychological blocks.

Erections on Demand

In the two decades before Viagra's release (c1980–1998), impotence was already deep in the process of being transformed once again, this time from a psychological condition to a physiological disorder. During this era the medical specialty of urology came to consider impotence a problem within its disciplinary purview and also developed new techniques and technologies for its diagnosis and treatment. This transformation occurred through a number of both coordinated and serendipitous events throughout the 1980s and 1990s. These culminated in 1998, with the release of Viagra to treat "erectile dysfunction," thereby solidifying its conceptualization as an organic condition, controllable through pharmacologic means.

Beginning in the 1980s, debates about the etiology of impotence began to intensify as urologists interested in developing new medical treatments sought to determine its underlying causes. Part of the insistence on the psychogenic causes of impotence in the 1960s and 1970s stemmed in large part from the fact that psychotherapeutic techniques were the only ones available to impotent men and their partners, with the marked exception of penile prostheses, which were available, yet not widely used throughout the second half of the twentieth century.[15] Understandings of underlying etiology of impotence began to shift due both to urologists' trial and error of new medical treatments and as failure rates of psychotherapeutic approaches began to emerge.[16] As is common in the production of scientific knowledge, new techniques and instruments can be crucial elements in clearing the path for an emergent science and scientific understanding.[17] As demonstrated throughout this chapter, ideas about the underlying physiological bases of impotence were then and continue to be coconstituted with new techniques for diagnosis and for treatment.

Although sex therapy dominated the discourse and clinical treatment of impotence throughout the 1970s and 1980s, during this time urologists also began to interest themselves in the problem from a medical and physiological perspective. Kinsey Institute director, John Bancroft, a psychiatrist by training and a prominent sexologist with over thirty years of experience in the field, explains this development through urologists' self-interest in maintaining their profession and livelihood. He stated:

> urology had something of a crisis. . . . Urology is a surgical specialty, and was being squeezed out by non-medical methods and management. So, urologists were turning to other types of problems to earn their living. Erectile dysfunction, or whatever you prefer to call it, was one that attracted their attention.[18]

Leonore Tiefer similarly linked urologists' attraction to sexual dysfunction as a subspecialty with the capacity to diversify their outpatient and inpatient services by clinically treating a patient population that was not in the strictest sense chronically "ill," "diseased," or likely to die from this condition.[19]

At first, as noted above, surgical solutions were offered to men with impotence by way of penile implants, inflatable prostheses of semirigid rods inserted into the penile shaft. These treatments were (and still are) invasive and not without significant risk of malfunction and infection. They were hardly considered ideal solutions by either physicians or patients. However, their availability contributed to the acceptability of remedies for "erectile dysfunction," which paved the way for Viagra's legitimacy and acceptability. Urologists thereby succeeded in establishing themselves as having the expertise and authority to treat impotence, and conducted an increasing amount of biomedical research on the physiological mechanisms of erections. Such treatments also made the etiology of erectile dysfunction irrelevant, since they worked regardless of the cause of impotence (e.g., psychological causes, iatrogenic effects of surgery or other treatments, or a medical condition). Mechanical treatments, which quite literally could give men erections on demand, also made sexual desire irrelevant for sexual performance. Implants created the ability for a man to have an erection irrespective of his state of desire or arousal.

Despite the seeming *irrelevance* of sexual desire for sexual activity, this conceptualization further *delineated* and *divided* processes of desire from processes of arousal. Instead of constructing the impotent man as desireless, discourses about implant use took male desire as already given. An erection in this context was therefore not a bodily response to a psychological state of being or feeling; rather, the mind and body were considered distinct and separate entities. This idea ran counter to psychological theories of impotence which posited that the means of treating impotence was to treat the mind first and then the body will follow. In fact, with the advent of new medical technologies such as implants, cause and effect were inverted. While before, the common wisdom was that fear, anxiety, depression, etc. could act on the body to produce impotence, now the emphasis was on the psychological distress *caused* by impotence. In turn, implants, through their mechanical means of repairing erections, could alleviate psychological problems: fix the body and the mind will follow. This switch marked a turning point in the reconfiguration of impotence as a "lifestyle" problem, wherein treating impotence could benefit and even enhance one's "quality of life."[20] Furthermore, the implication was that impotence itself is a mechanical problem in need of a mechanical solution.

The Organicization of Impotence:
From Impotence to Erectile Dysfunction

Throughout the 1980s and 1990s, urologists continued to study the physiology of erections and to search for pharmacological treatments. Over time, the notion that "over 90 percent of impotence is psychologically based" began to shift to a more equitable distribution: eventually, urologists and mass media claimed that about half of all impotence cases had an *organic* basis.[21] A number of marker events signaled this turning tide, as urologists begin to usurp what was previously the territory of sex therapists in both treating and defining impotence. In 1980, endocrinologist Richard Spark and his colleagues published an article in *JAMA* entitled, "Impotence Is Not Always Psychogenic"[22] He argued that, contrary to the claim that most cases of impotence are psychogenic, there are likely many cases of impotence that have their origins in hormonal abnormalities. He reported the results of a study of a sample of 105 "impotent" patients who had each had their serum testosterone levels checked;[23] 35% had hormonal abnormalities.[24] This article's title became something of a battle cry for clinicians and medical researchers in the following decades.

A second event reached legendary proportions in the sexual dysfunction lore. It occurred in 1983 at the American Urological Association meetings when Giles Brindley, a urologist, injected his own penis with phenoxybenzamine,[25] a pharmacologic solution that gave him an erection. He then presented his erect penis as specimen, walking the aisles of the conference room letting others inspect and touch it. This was often referred to as the watershed moment in the shift from impotence as a psychological to a physiological condition. In an early case of "reverse engineering" in this field, producing an erection by chemical means cast light on the physiological mechanisms of erections, codifying a *lack of* erections (that is, impotence) as a physiological condition. In other words, if an erection could be biochemically induced, then when one is unable to have an erection, there must be a biochemical problem. Because the penile injection used by Brindley was a "vasoactive" substance, meaning that it opened up blood vessels allowing greater blood flow, erections were rein-terpreted as a function of blood flow. Like implants, injections such as these could produce erections regardless of the etiology of the impotence, and regardless of whether or not a man had any difficulty getting an erec-tion "on his own."[26] Working within the same paradigm as penile implants, penile injections induced erections irrespective of a man's "feel-ings" of desire or arousal.

These innovations were considered by some scholars as a confirmation and reification of the phallocentrism of male sexuality, wherein construc-tions of male sexuality focused exclusively on penile erection for sexual pleasure.[27] However, while appreciated by clinicians and male users for their "reliability,"[28] penile injections, implants, and vacuum pumps were not considered ideal treatments by patients or urologists precisely because of the "artificialness" of the erections produced in this manner.[29] As dis-cussed later, Viagra provided a nice foil to earlier innovations, in that it was capable of producing an erection through a noninvasive pill that cre-ated erections through "natural" processes.

Throughout the 1980s and 1990s, various techniques and devices were developed and implemented in order to differentiate diagnoses of organic versus psychogenic impotence. A medical device called the RigiScan was developed by Timm Medical Technologies, Inc., to measure penile rigidity. Still widely used today, it consists of a recording device and two loops which are placed around the base of the penis and the tip of the penis. The loops tighten periodically (usually every 15 or 30 seconds) and a recording of the rigidity of the penis is calculated. The device can be operated by patients in the laboratory or in at-home settings. It is often used to mea-sure "nocturnal tumescence," meaning that a man wears the device while he is sleeping, where it measures the degree of the rigidity of the penis at 30-second intervals over the course of the night. It is thought that men who are able to get nocturnal erections likely have impotence with a psy-chogenic origin because these men do not seem to have physiological difficulties.

Through such research, the demarcation between "organic" and "psy-chogenic" etiology of impotence became more distinct during this time period, as it was thought that all erections (i.e., those that emerge from sexual stimuli and those that do not) were the same. That is, if a man was capable of nocturnal erections, then he should be physiologically capable of erections for sexual activity. Once again, the question of desire was elided or relegated to the "psychological" sphere.

A second diagnostic technique developed during this time was duplex ultrasonography.[30] As erectile function came to be understood through theories of vascularization and blood flow, the duplex ultrasound was used to evaluate the status of penile arteries. Ultrasound could detect arterial blockages that might prevent erectile functioning. Since its introduction in 1985, duplex ultrasonography has become the most commonly used clinical test for ascertaining the vascular status of patients with erectile dysfunction (ED), in order to differentially diagnose it.[31]

During the 1980s, while working on the physiology of erections and potential treatments, a group of urologists also made a concerted effort to reconstruct impotence as a medical rather than psychological condition. As detailed by Tiefer, in 1982, this group of urologists held an informal meeting which resulted in the formation of the International Society for Impotence Research.[32] The first World Meeting on Impotence was held in 1984. A few years later, three urologists published an overview of this "emergent" field in the *New England Journal of Medicine*.[33] In 1989, the International Society for Impotence Research began publishing its own medical journal, *The International Journal of Impotence Research* whose editorial board primarily consisted (and continues to consist) of urologists.

In 1992, this shift of impotence from a psychological to a biomedical disorder was stabilized and codified by a "consensus development conference" sponsored by the National Institutes of Health. This also marked the beginnings of the transformation of terminology from impotence to "erectile dysfunction" (ED). A number of "experts" in the field of impotence research gathered in Bethesda for a day and a half public session to present their own work in areas relevant to impotence research, followed by an open public discussion session.[34] A separate panel then met behind closed doors to develop a consensus statement based on the previous days' discussions. The planning committee for the conference consisted entirely of medical doctors and NIH employees. The experts who presented were mostly (75%) physicians.

Reports of the consensus conference appeared in at least two venues. The first was in the *International Journal of Impotence Research* in 1992, and the second later appeared in *JAMA*.[35] Although the *JAMA* document was entitled "Impotence," the first paragraph said that although *impotence* has been traditionally used to signify the inability of the male to attain and maintain erection of the penis, this use had led to confusion within scientific investigations because of its lack of specificity. It then continued:

"this, together with its pejorative implications, suggests that the more precise term *erectile dysfunction* be used instead."[36] The article defined erectile dysfunction as "the inability to attain and/or maintain penile erection sufficient for satisfactory sexual performance."[37] The renaming process codified men's sexual problems as medical conditions, encouraging further medical research to understand male sexual function. The chronology of events here indicates that this was not simply a terminological change, but rather an indication of medical claims-making about the organic and physiological etiology of the disorder itself.

In retrospect, this seems an obvious and purposeful way to reframe impotence within medical parlance, through focusing on the functionality of a specific organ. However, surprisingly this change was first recommended by psychologists themselves. Leonore Tiefer, a psychologist by training and a critic of "the medicalization of male sexuality,"[38] in 1994 claimed that it was *psychologists* who preferred the terms "erectile disorder" or "erectile dysfunction" and urologists who promoted their own claims through the continual use of the word *impotence*.[39] The term "dysfunction" itself, in existence since at least the 1970s as in the terminology of "sexual dysfunction," was initially used by sex therapists.[40] In fact, at the 1992 consensus development conference, Tiefer, speaking on "nomenclature," was the one to suggest that erectile dysfunction replace the term impotence, which she felt was pejorative and confusing.[41] She advocated use of the term *erectile dysfunction* as a strategy for the *de*-medicalization and de-stigmatization of impotence. Given Tiefer's stance on the dangers of medicalization, it was ironic that she provided nomenclature that later served to help medicalize the condition. In ways that appear predictable in retrospect, erectile dysfunction, as later heavily promoted by Pfizer vis-à-vis Viagra, in many ways legitimated impotence as a biomedical condition, wresting it away from its much older psychoanalytic and psychological associations. The new term capitalized on its ability to de-stigmatize and sanitize the condition. Furthermore, it focused attention on erections themselves as the source of the problem, rather than, as most psychologists would prefer, a less local and more holistic dysfunction around sexuality broadly conceived.

Accompanying the organicization of erectile dysfunction, the consensus statement also concretized and institutionalized the emergent idea that ED is not a "normal" occurrence as men age. In a section on "prevention," the report said that "although erectile dysfunction increases progressively with age, it is not an inevitable consequence of aging."[42] As in the broader discourse of aging in contemporary biomedicine, aging itself needed not be accompanied by any disease or even limitations,[43] only an increasing coalescence of risk factors. With ED, such risk factors (which often "accompany" aging, but are not integral to it) included diabetes mellitus, hypertension, vascular disease, high cholesterol, depression, and alcohol ingestion.[44]

In some ways, then, our notion of impotence has come full circle. Our conception of impotence at the beginning of the twentieth century was as an inevitable organic process that occurs as men age. After the shift to understanding impotence as a psychological condition, once again we have returned to reconceptualizing it as a physiological and once again organic process. Only this time around, erectile dysfunction has been decoupled from the natural aging process and thereby pathologized and subsequently medicalized. Although the process of *pathologization* and *medicalization* are often related, they are not necessarily coterminous or coincident. In the case of erectile dysfunction, impotence first became pathologized through psychological interventions and changing norms about the benefits and prevalence of sexual activity in older men. The idea that men continued to be sexual as they age made those who were not or could not be abnormal, thereby making impotence pathological. It wasn't until later, in the 1990s, that we witnessed the complete biomedicalization of impotence, whereby not only was it considered pathological, but also as a physiological problem in need of a biomedical solution.

The NIH Consensus Statement was both a result of and instigator for medical attention to the problem of erectile dysfunction. The work conducted throughout the 1990s on the physiology of erectile function is now reflected upon by urologists as a foundational example of the ways in which medical research can advance our knowledge about the inner workings of the human body. For example, one urologist began an article on Viagra's clinical success for treating erectile dysfunction as follows: "[i]n the past 15 y[ears], erectile dysfunction (ED) has witnessed an enormous growth in scientific interest, which has been translated impressively to the development of several treatment options."[45] A urologist who began studying erectile function in the early 1980s said similarly:

> You know, over the last fifteen to twenty years I think . . . researchers made a lot of progress in terms of: understanding what is a normal physiology; what happens during an erection; what is the pathophysiology for why impotence occurs; why diabetes causes impotence; why high cholesterol

causes impotence. You know I think there has been a lot of progress made in better understanding.[46]

With continued focus on the physiology of the penis, attention shifted to studying its functioning as an organ, in much the same way that a medical researcher might study the kidney or even the heart; it became devoid of most of the remnants of the psychological components of erections.

The Neurochemistry of Erections: The Explanatory Power of Nitric Oxide

Research conducted during the 1990s produced a theory of the physiology of erection that was dependent upon a complex interplay between vascular and neurological events. As described in the NIH Consensus Statement, erections are most often initiated by a central nervous system response to psychogenic stimuli (e.g., "desire," perception, etc.), which then activates the sympathetic and parasympathetic innervation of the penis. The sympathetic aspects of the sexual arousal process had long been theorized. These are the observable aspects of arousal detailed by Masters and Johnson, including heart rate increases, contraction of voluntary muscles, and so on.[47] The parasympathetic aspects of the arousal response were newer elements in the theory and ran counterintuitive to previous notions of sexual arousal. Sexual arousal is usually described as an excitatory process, one that should then be mediated by sympathetic nervous system responses. However, in the 1980s, it was found that parasympathetic input, which generally speaking has a relaxation effect, is also required for sexual arousal to occur.[48] This relaxation occurs at the level of the smooth muscle of the corpus cavernosa of the penis. The corpus cavernosa are the two parallel chambers of the penis that run alongside the urethra (see Figure 1). When the smooth muscle relaxes or vasodilates, more room is created in the corpus cavernosa for blood to flow into the chambers, causing an erection. However, the mechanisms by which this reaction occurred were not understood at the time.

In 1992, an article published in the *New England Journal of Medicine* reported the results of a study that demonstrated the role of nitric oxide as a mediator of the smooth muscle relaxation of the corpus cavernosa of the penis.[49] The lead author of the study, Jacob Rajfer, is a urologist, and the last author listed, Louis Ignarro, is a pharmacologist.[50] Both were (and

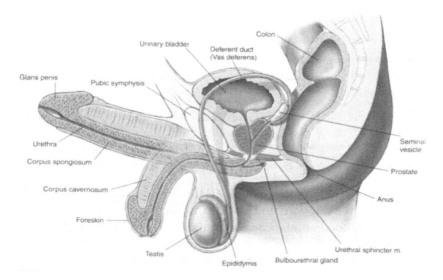

Figure 1.

remain) professors at UCLA. Ignarro had spent the previous 15 years studying the potential effects of nitric oxide on cardiovascular tissue, publishing his first article in this area in 1979. Along with two other American researchers at different academic institutions, he was awarded the Nobel Prize in Medicine in 1998 (the same year as Viagra's release) for their work in identifying nitric oxide as a messenger signal of the cardiovascular system, inducing relaxation of the smooth muscle of blood vessels. In the 1990s, he and Rajfer conducted some of the research identifying nitric oxide as the neurotransmitter responsible for smooth muscle dilation in the penis, enabling erections. Nitric oxide is released by both the cavernosal nerves and the epithelial cells of the corpus cavernosum. Up to this time, a number of neurotransmitters were considered responsible for creating erections, including both vasoactive intestinal polypeptide (VIP) and prostaglandin E_1. Prostaglandin E_1 does play a role in inducing erections; it is the active compound in penile injections. However, most penile injection drugs were based not on the outcome of research on the physiology of erection but through the trial-and-error efforts of empirical work.[51]

The research conducted by this group as well as numerous others using in vitro samples of animals and humans resulted in an understanding of erectile response that can be diagrammed as follows:[52]

A Physiological Explanation of Erections

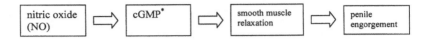

* cGMP stands for cyclic guanosine monophosphate, a messenger which acts on certain proteins to produce smooth muscle relaxation.

Figure 2.

Finding the mechanism that induces erections lent further credence to the idea that erectile response, and male sexual arousal more generally, could be best understood as a biochemical, mechanical, and physiological phenomenon. In fact, the most promising means of developing new drugs to create seemingly "natural" erections with fewer side effects was through investigation of the physiological pathways of erections.

Viagra and Notions of Serendipity in Science

The tale often told of Viagra's discovery relies on a discourse of serendipity. In this story, Pfizer in the early 1990s had developed sildenafil citrate (the chemical name for what eventually became Viagra) as a potential treatment for angina pectoris. During the clinical trials, in which men were taking high doses of sildenafil on a daily basis, some reported that they were having "better" erections.[53] Pfizer, linking this development to emergent physiological theories of penile blood flow, immediately began clinical research on sildenafil as a treatment for erectile dysfunction.

Capitalizing upon this knowledge, the clinical trials on sildenafil were subsequently modified to test for the treatment of erectile dysfunction in men. The clinical trial process for what became Viagra progressed with alacrity. While a drug usually spends close to 10 years in clinical research development before it is approved by the U.S. Food and Drug Administration, Viagra sped through the trial process in less than five years. This was due in part to the fact that the drug compound had already been developed, which again speaks to Pfizer's good fortune in finding a different popular use for it. In addition, much of the proof of principle of sildenafil—that is, a "theory" about how the drug works—was already well established by the ongoing research on the cGMP pathway of erections. Pfizer simply needed to fit sildenafil into the model established by academic scientists over the previous years.

It was soon discovered that sildenafil was a phosphodiesterade type 5 (PDE5) inhibitor. Within the cGMP pathway outlined above, PDE5 *breaks down* cGMP. In other words, after cGMP is released causing muscle relaxation, PDE5 is responsible for metabolizing the cGMP, returning the smooth muscle back to its tonic, contracted state. Sildenafil, blocks the activity of PDE5, thereby maintaining high levels of cGMP, maintaining relaxation of the smooth muscle of the penile chambers and thereby allowing for greater blood flow into the penis over a longer than usual period of time. Sildenafil in effect *enhances* the effect of the cGMP pathway for erectile function. Figure 3 diagrams this process.

Using corpus cavernosal tissue from men who had undergone surgery for penile prostheses, a study conducted by Pfizer researchers in the United Kingdom and urologists at a hospital in Bristol, England demonstrated that sildenafil does in fact interfere with the cGMP pathway, enhancing the nitric oxide effects of muscle relaxation.[54] A similar study was conducted on rabbit corpus cavernosal tissue with the same findings.[55]

The Phase I and Phase II clinical trial studies of Viagra progressed with similar ease. Phase I is a clinical trial on a few persons to determine the safety of a new drug and its dosage or toxicity limits. Phase II of a clinical drug trial is when a drug is tested on human subjects to evaluate its

Viagra's Mechanism of Action

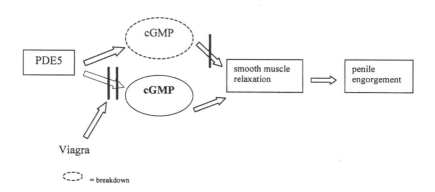

Figure 3.

efficacy and side effects. Phase III is a large clinical trial of a drug that in phase I and phase II has been shown to be efficacious with tolerable side effects; after successful conclusion of phase III clinical trials, a company applies for formal approval from the U.S. Food and Drug Administration (FDA).[56] By April 1996, Pfizer had completed the Phase I and Phase II trials of Viagra and even had a trademark name. An article appeared in the *International Journal of Impotence Research* in 1996 describing results from five different studies—one in vitro test of Viagra's effects on human corpus cavernosal tissue; three pharmacokinetic studies where men were given Viagra to have their blood and urine assayed; and one clinical "laboratory" study where men were given Viagra in a controlled environment to test the efficacy of the drug.[57] The laboratory study, or Phase II of the clinical drug trial process, described in this article was one of the only published accounts of Viagra taken in the laboratory. This study was a double-blind placebo controlled study in which 12 men came into the clinic four times. On each occasion they were given either a placebo or one of three doses of Viagra. They were then asked to choose from a selection of sexually explicit videos and magazines which they began viewing 30 minutes postdose. Efficacy of the drug was evaluated by measurements taken by the RigiScan which measures penile rigidity at the tip and base of the penis. Erectile function is based on ideas about what is necessary for penetrative and heterosexual sex. Mean duration of rigidity of greater than 60% at the base and the tip was analyzed. Sixty percent rigidity is deemed to be the minimum pressure deemed necessary for adequate penetration of the vagina by the penis.[58] This heteronormative imperative became embedded in the clinical trial of the drug itself, as success was measured by what would sufficiently enable heterosexual intercourse.[59]

Phase III studies were conducted across the United States (and elsewhere). Men were given either Viagra or a placebo and asked to take a pill at home one hour before sexual activity. In total, 861 men were studied at 37 health centers across the United States.[60] Just two months after Viagra's FDA approval for clinical use, an article reporting the results of the Phase III trials in the United States appeared in the *New England Journal of Medicine* in May.[61]

This "accidental discovery," while a significant component of Viagra's development, does not sufficiently consider urologists' concerted efforts to search for pharmacologic treatments throughout the 1980s and 1990s, nor does it recognize the contingency inherent in these developments. Relying on "serendipity" is a way of maintaining a notion that technologies, scientific facts, and scientific knowledge developments, more generally, are simply out there in the world, waiting to be discovered, rather than social and cultural productions in their own right. Many reports of scientific discovery include a component of serendipity, the fortuitous happenstance of trying the right chemical agent at the right time. Serendipity is often used unproblematically in histories and social studies of science. In addition, in scientists' own retelling of their work, serendipity often plays a role in the reconstruction of the story of discovery.[62]

Relying on notions of serendipity for the explanation of scientific development can also be read as a verification of the "great man" theory of science, where instead of looking at the social and cultural contingencies of a scientific development, one relies on an individual's (or in this case, a pharmaceutical company's) "genius" for recognizing and identifying what might be otherwise construed as "luck." Serendipity, therefore, often refers not to luck or an acknowledgment of contingency, but rather to a type of *domestication* or *harnessing* of luck through an individual's genius, fortitude, savvy, etc. However, in the case of Pfizer's serendipitous discovery of Viagra, the "discovery" itself only made sense (in other words, it was serendipitous rather than meaningless) because what they found *could* be explained by emergent theories within cardiovascular and urological research. Therefore, serendipity did play a role in Viagra's development, less for the scientific study of erections, but more for Pfizer, the pharmaceutical company that first developed the drug.

The history of the development of Viagra is in many ways typical of other pharmaceutical technologies. While academic and other research scientists often conduct basic scientific research, it is private biochemical and pharmaceutical companies that apply this research to create profitable products.[63] Sildenafil did not seem to be effective for treatment of angina pectoris, but, because of the research conducted on the physiological mechanisms of erections, sildenafil could be rehabilitated as a treatment for erectile dysfunction. Once male clinical trial subjects reported that they were experiencing better erections with sildenafil, Pfizer was able to capitalize quickly on this knowledge and on the concurrent basic science research already conducted on the physiology of erections to develop a successful drug.

Perhaps more important than the scientific developments leading up to Viagra's "discovery" were the research networks, organizations, and overall transformation of biomedicine's conceptualization of the nature of impotence constructed in the two decades prior to the compound that came to be known as Viagra. Not only did the concerted efforts of urologists and others contribute to the science of Viagra, but moreover they contributed to a shift in the social and cultural climate. This made Viagra a credible, legitimate, and successful commodity both for medical practitioners and consumers more broadly for treating what is now known as erectile dysfunction. Developments throughout the last quarter of the twentieth century reconstituted impotence as a physiological phenomenon and a legitimate medical condition separate and apart from the normal process of aging. It was this environment that enabled the emergence of Viagra and foretold its success.

In sum, the scientific and cultural understandings of "erectile dysfunction" over the 25 years prior to Viagra's release laid the groundwork for its subsequent acceptability and success. First, the psychologization of impotence in the 1960s and 1970s pathologized impotence, such that it was no longer considered a component of normal aging, but as a dysfunction in need of treatment. The organicization of impotence beginning in the 1980s relocated impotence from a problem of the mind to a problem of the body, focusing attention on the penis as an "organ." When taken together, these ideas transformed impotence into a "lifestyle" condition. If impotence was neither inevitable for aging men nor indicative of larger psychological problems, it became simply a condition to be treated, not because it was life-threatening, but rather because it is life-limiting. The medicalization of the problem as "erectile dysfunction" legitimated it as one in need of a medical solution. Erectile dysfunction, then, needed to capitalize on both its status as a medical condition in need of treatment and a lifestyle condition that deserved treatment because of the hindrances it placed on one's quality of life.

Although the discovery of sildenafil's effects on aiding erection was indeed unintentional and serendipitous, the concerted effort on the part of researchers and Pfizer in realizing Viagra as a viable commercial product was anything but accidental. Despite the seemingly unexpected arrival of Viagra on the pharmaceutical market, its success and availability was contingent upon the years of dedicated research and effort into rethinking of impotence as a medical condition. Without the medicalization of impotence, the FDA could not have approved Viagra as a prescription drug to treat "erectile dysfunction." Urologists' intentions to claim jurisdictional ownership of impotence and the intense focus on uncovering the physiology of erections led not only a medicalization, but also a molecularization of impotence. This was instrumental in paving the way for Viagra's eventual breakthrough. Furthermore, understanding erectile dysfunction as a medical condition lent Viagra an air of legitimacy that catapulted it into the blockbuster drug strata (with the help of private and Medicare insurance coverage).

Rather than understanding Viagra's release as a singular and solitary event, this chapter has attempted to place it in historical and social context. This is important not only for understanding the particulars of Viagra, but also for understanding technological developments more broadly. If we are to understand technological innovation, we must realize that technologies cannot be separated from the social, historical, or scientific context from which they emerge.

Notes

1. For an analysis of the deployment of "novelty" in biomedicine, see Nicholas B. King, "Infectious Disease in a World of Goods" (PhD dissertation, Harvard University, 2001).

2. Adele E. Clarke, Janet K. Shim, Laura Mamo, Jennifer Ruth Fosket, and Jennifer R. Fishman, "Biomedicalization: Technoscientific Transformations of Health, Illness, and U.S. Biomedicine," *American Sociological Review* 68 (2003), 161–194.

3. Barbara L. Marshall and Stephen Katz, "Forever Functional: Sexual Fitness and the Ageing Male Body," *Body and Society* 8 (2002), 43–70.

4. Brown-Séquard may be most well known for his early work with identifying "sex hormones" through his now famous experiments using animal gland injections in other animals and humans. See Adele E. Clarke, *Disciplining Reproduction: Modernity, American Life Sciences, and the Problems of Sex* (Berkeley, 1998); Nelly Oudshoorn, *Beyond the Natural Body: An Archeology of Sex Hormones* (New York, 1994).

5. Ibid. Marshall and Katz, "Forever Functional," 52.

6. As cited in Ibid. Marshall and Katz, "Forever Functional."

7. Ibid. Marshall and Katz, "Forever Functional," 52.

8. Alfred C. Kinsey et al., *Sexual Behavior in the Human Male* (New York, 1948); William H. Masters and Virginia Johnson, *Human Sexual Inadequacy* (Boston, 1970); William H. Masters and Virginia E. Johnson, *Human Sexual Response* (Boston, 1966).

9. Leonore Tiefer, *Sex Is Not a Natural Act and Other Essays* (Boulder, 1995).

10. Francois Ewald, "Norms, Discipline, and the Law," *Representations* 30 (1990), 138–161.

11. By sexology, I refer here not only to the discipline dedicated to the study of sex, but also to the notion that sexuality is an area worthy of *scientific* pursuit.

12. Sheri Hite, *The Hite Report* (New York, 1976).

13. John Bancroft, "Erectile Impotence: Psyche or soma?" *International Journal of Andrology* 5 (1982), 353–355.

14. See, for example, Helen Singer Kaplan, *The New Sex Therapy: Active Treatments of Dysfunctions* (New York, 1974); Richard F. Spark, Robert A. White, and Peter B. Connolly, "Impotence Is Not Always Psychogenic," *Journal of the American Medical Association* 243 (1980); and Robert N. Butler and Mitzi L. Davis, *Love and Sex After 60,* Revised ed. (New York, 1988).

15. Leonore Tiefer, "In Pursuit of the Perfect Penis: The Medicalization of Male Sexuality," *American Behavioral Scientist* 29 (1986), 579–599. Tiefer estimates that "hundreds of thousands of men had received implants" by the mid-1980s, which seems miniscule when compared to Pfizer's statistics of the 20 million men, who had had Viagra prescriptions for them by the end of 2002. Yet, Tiefer also points out the difficulty in estimating the number of penile implant operations conducted because they are implanted primarily by private practitioners.

16. See, for example, Ibid. Spark, White, and Connolly, "Impotence Is Not Always Psychogenic."

17. Georges Canguilhem, *Machine and Organism* (New York, 1992); Georges Canguilhem, *On the Normal and the Pathological,* trans. Carolyn R. Fawcett (Dordrecht, 1978); Bruno Latour, *Science in Action: How to Follow Scientists and Engineers Through Society* (Cambridge, MA, 1987).

18. From interview transcript R10, 10/1/01, lines 37–45.

19. Leonore Tiefer, "The Medicalization of Impotence—Normalizing Phallocentrism," *Gender & Society* 8 (1994), 363–377.

20. In studies to test the effectiveness of erectile dysfunction treatments, "quality of life" became a meaningful outcome variable in and of itself. See, for example, Stanley E. Althof, "Quality of Life and Erectile Dysfunction," *Urology* 59 (2002), 803–810; Mark S. Litwin, Robert J. Nied, and Nasreen Dhanani, "Health-Related Quality of Life in Men with Erectile Dysfunction," *Journal of General Internal Medicine* 13 (1998), 159–169; S. N. Seidman S. P. Roose, M. A. Menza, R. Shabsigh, and R. C. Rosen, "Sildenafil improved erectile dysfunction and quality of life in men with comorbid mild-to-moderate depression," *ACP Journal Club* 137 (2002), 21; Jean M.B. Woodward, Steven L. Hass, and Paul J. Woodward, "Reliability and Validity of the Sexual Life Quality Questionnaire," *Quality of Life Research* 11 (2002), 365–377.

21. Ibid. Tiefer, "In Pursuit of the Perfect Penis."

22. Ibid. Spark, White, and Connolly, "Impotence."

23. In this article, impotence was determined according to the same standards that Masters and Johnson used at the time: a man was defined as impotent if

"erectile failure" occurred in more than 25% of the attempts at intercourse. Ibid. Spark, White, and Connolly, "Impotence."

24. The methodological problems with this article are numerous, although not particularly relevant to the discussion at hand. Because of the article's prominent placement in the medical literature and its decisive sentence title, the argument carried much weight in propelling a reconfiguration of impotence as a medical problem.

25. Phenoxybenzamine is a drug classified as an antihypertensive and alpha adrenergic blocker. The drug is currently available in capsules only, not for injection.

26. See, for example, H. Ghanem T. Sherif, T. Abdel-Gawad, T. Asaad, "Short term use of intracavernous vasoactive drugs in the treatment of persistent psychogenic erectile dysfunction," *International Journal of Impotence Research* 10 (1998), 211–214.

27. Ibid. Tiefer, "The Medicalization of Impotence."

28. L. Incrocci, W. C. J. Hop, and A. K. Slob, "Visual erotic and vibrotactile stimulation and intracavernous injection in screening men with erectile dysfunction: a 3 year experience with 406 cases," *International Journal of Impotence Research* 8 (1996), 227–232.

29. D. G. Hatzichristou, "Sildenafil Citrate: Lessons Learned from 3 Years of Clinical Experience," *International Journal of Impotence Research* 14 (2002), S42–S52; Somboon Leungwattanakji, Vincent Flynn, and Wayne J. G. Hellstrom, "Intracavernosal Injection and Intraurethral Therapy for Erectile Dysfunction," *Urologic Clinics of North America* 28 (2001), 343–354; D. L. Rowland, H. S. M. Boedhoe, G. Dohle, A. K. Slob, "Intracavernosal self-injection therapy in men with erectile dysfunction: satisfaction and attribution in 119 patients," *International Journal of Impotence Research* 11 (1999), 145–151.

30. For a history and analysis of duplex ultrasonography, see Annemarie Mol, *The Body Multiple: Ontology in Medical Practice* (Durham, NC, 2002).

31. T. G. W. Steel, H. van Langen, H. Wijktra, E. J. Meuleman, "Penile Duplex Pharmaco-Ultrasonography Revisited: Revalidation of the Parameters of the Cavernous Arterial Response," *Journal of Urology* 169 (2003), 216–220.

32. Ibid. Tiefer, "The Medicalization of Impotence."

33. R. J. Krane, I. Goldstein, and I. S. DeTejada, "Impotence," *New England Journal of Medicine* 321 (1989), 1649–1659.

34. NIH Consensus Development Panel on Impotence, "Impotence: NIH Consensus Conference," *Journal of the American Medical Association* 270 (1993).

35. National Institute of Health, "Consensus Development Conference Statement on Impotence," *International Journal of Impotence Research* 5 (1992). Ibid. NIH Consensus Development Panel on Impotence, "Impotence."

36. Ibid. NIH Consensus Development Panel on Impotence," "Impotence," 83, emphasis in original.

37. Ibid. NIH Consensus Development Panel on Impotence, "Impotence," 89.

38. Ibid. Tiefer, "In Pursuit of the Perfect Penis."

39. Ibid. Tiefer, "The Medicalization of Impotence."

40. See, for example, J. LoPicollo, "Direct Treatment of Sexual Dysfunction," in *Handbook of Sex Therapy*, ed. J. LoPicollo and P. LoPicollo (New York, 1978).

41. Ibid. Tiefer, *Sex Is Not a Natural Act*.

42. Ibid. NIH Consensus Development Panel on Impotence, "Impotence," 85.

43. Michael Bury, "Health, Ageing, and the Lifecourse," in *Health, Medicine and Society: Key Theories, Future Agendas*, ed. Simon Williams, Jonathan Gabe, and Michael Calnan (London, 2000), 87–105; Carl I. Cohen, "Old Age, Gender and Physical Activity: The Biomedicalization of Aging," *Journal of Sport History* 18 (1991), 64–80; Carroll L. Estes and Elizabeth A. Binney, "The Biomedicalization of Aging: Dangers and Dilemmas," *The Gerontologist* 29 (1989), 587–596; Mike Featherstone and Mike Hepworth, "The Mask of Ageing and the Postmodern Life Course," in *The Body: Social Process and Cultural Theory*, ed. Mike Featherstone, Mike Hepworth, and Bryan S. Turner (London, 1991), 371–189; Ibid. Marshall and Katz, "Forever Functional."

44. Ibid. National Institute of Health, "Consensus Statement on Impotence," 85.

45. Ibid. Hatzichristou, "Sildenafil Citrate," S43.

46. From interview transcript R20, 12/28/01, lines 74–82.

47. Ibid. Masters and Johnson, *Human Sexual Response*.

48. T. F. Lue, "Hemodynamics of Erection in the Monkey," *Journal of Urology* 130 (1983), 1237–1241.

49. J. Rajfer, W. J. Aronson, P. A. Bush, F. J. Dorey, and L. J. Ignarro, "Nitric Oxide as a Mediator of Relaxation of the Corpus Cavernosum in Response to Noradrenergic, Noncholinergic Neurotransmission," *New England Journal of Medicine* 326 (1992), 90–94.

50. Unlike social science journal articles, in medical journal articles it is often the case that the principal investigator of a study is listed last.

51. Christian G. Stief, Stefan Uckert, Armin J. Becker, Michael C. Truss, and Udo Jonas, "The Effect of the Specific Phosphodiesterase (PDE) Inhibitors on Human and Rabbit Cavernous Tissue in Vitro and in Vivo," *Journal of Urology* 159 (1998), 1390–1393.

52. See, for example, A. L. Burnett, "Role of Nitric Oxide in the Physiology of Erection," *Biological Reproduction* 52 (1995), 485; A. L. Burnett, C. J. Lowenstein, D. S. Bredt, T. S. K. Chang, and S. H. Snyder, "Nitric Oxide: A Physiologic Mediator of Penile Erection," *Science* 257 (1992); 41; P. A. Bush, W. J. Aronson, M. B. Georgette, J. Rajfer, and L. J. Ignarro, "Nitric Oxide Is a Potent Relaxant of Human and Rabbit Corpus Cavernosum," *Journal of Urology* 147 (1992), 1650; F. Trigo-Rocha, W. J. Aronson, M. Hohenfeller, L. J. Ignarro, J. Rajfer, and T. F. Lue, "Nitric Oxide and cGMP: Mediators of Pelvic Nerve-Stimulated Erection in

Dogs," *American Journal of Physiology* 264 (1993), H419–H422; A.W. Zorgniotti and E. F. Lizza, "Effect of Large Doses of the Nitric Oxide Precursor, L-arginine, on Erectile Dysfunction," *International Journal of Impotence Research* 6 (1994), 33–36.

53. Although not defined in any of the reports of this phenomenon, it seems that "better" in this case refers to erections that were harder and lasted longer. Only men were enrolled in this study. Because this was not a government-funded study, there was no mandate that Pfizer include women in the research. This is an interesting point because we do not know if *women* would have reported any improvement in their sexual arousal response. Jacob Rajfer, "From the Lab to the Clinic," *Journal of Urology* 159 (1998), 1792; John Leland et al., "A Pill for Impotence?" *Newsweek,* November 17, 1997, 62–68, and interview data 10/21/01.

54. Stephen A. Ballard, Clive J. Gingell, Kim Tang, Leigh A. Turner, Mary E. Price, and Alasdair M. Naylor, "Effects of Sildenafil on the Relaxation of Human Corpus Cavernosum Tissue in Vitro and on the Activities of Cyclic Nucleotide Phosphodiesterase Isozymes," *Journal of Urology* 159 (1998), 2164–2171.

55. Alex T. Chuang, John D. Strauss, Richard A. Murphy, and William D. Steers, "Sildenafil, a Type-5 cGMP Phosphodiesterase Inhibitor, Specifically Amplifies Endogenous cGMP-Dependent Relaxation in Rabbit Corpus Cavernosum Smooth Muscle in Vitro," *Journal of Urology* 160 (1998), 257–261.

56. Harry M. Marks, *The Progress of Experiment: Science and Therapeutic Reform in the United States, 1900–1990* (New York, 1997).

57. Mitradev Booleil, Michael J. Allen, Stephen A. Ballard, Sam Gepi-Attee, Gary J. Muirhead, Alasdair M. Naylor, Ian H. Osterloh, and Clive Gingell, "Sildenafil: an orally active type 5 cyclic GMP-specific phosphodiesterase inhibitor for the treatment of penile erectile dysfunction," *International Journal of Impotence Research* 8 (1996), 47–52.

58. R. P. Allen, J. K. Smolev, R. M. Engel, and C. B. Brendler, "Comparison of RigiScan and Formal Nocturnal Penile Tumescence Testing in the Evaluation of Erectile Rigidity," *Journal of Urology* 149 (1993), 1265–1268; Ian Eardley, Peter Ellis, Mitradev Booleil, and Maria Wulff, "Onset and Duration of Action of Sildenafil Citrate fro the treatment of Erectile Dysfunction," *British Journal of Clinical Pharmacology* 53 (2002), 61S–65S; I. Goldstein, S. Auerbach, H. Padma-Nathan, J. Rajfer, W. Fitch III, L. Schmitt, and Alprostadil Alfadex Group, "Axial penile rigidity as primary efficacy outcome during multi-institutional dose titration clinical trials with alprostadil alfadex in patients with erectile dysfunction," *International Journal of Impotence Research* 12 (2000), 205–211.

59. This was also the case throughout the Phase III at-home clinical trials of Viagra.

60. This is the figure reported in the *New England Journal of Medicine* article published two months after Viagra's approval by the FDA. The pamphlet distributed with Viagra prescriptions says that Viagra was distributed to over 3000 men

worldwide before its approval in the United States. This larger figure includes other study designs than the one presented here and studies conducted outside of the United States. The key protocol, subject inclusion criteria, and outcome variables remained the same across all the studies. Irwin Goldstein et al., "Oral Sildenafil in the Treatment of Erectile Dysfunction," *New England Journal of Medicine* 338 (1998), 1397–1404. Pfizer Labs, "VIAGRA (Sildenafil citrate): Confidence in Treating ED," (New York, 2000); Pfizer Labs, "What Every Man (and Woman) Should Know: VIAGRA Pamphlet" (New York, 1998).

61. Ibid. Goldstein et al., "Oral Sildenafil in the Treatment of Erectile Dysfunction."

62. For example, Ignarro, mentioned earlier for his work on understanding the role of nitric oxide, too referred to a serendipitous occurrence in which nitric oxide was accidentally used in the laboratory (fieldnotes 11/16/2000). In an unrelated example, an interview with Richard Vale, professor and vice-chair of the Department of Cellular and Molecular Pharmacology at UCSF, appeared in the UCSF magazine *Newsbreak*, in which Vale emphasizes the importance of serendipity in the scientific discovery process. Jeffrey Norris, "On the Road with Molecular Motors," *Newsbreak: News for the UCSF Campus Community*, May 9, 2003, 1, 4.

63. See, for example, Jean-Paul Gaudilliere and Ilana Lowy, eds., *The Invisible Industrialist: Manufacturers and the Construction of Scientific Knowledge* (London, 1998).

About the Contributors

ROBERT BUD is Principal Curator of Medicine and Manager of Electronic Access at the Science Museum in London, where he was project director of the award-winning www.ingenious.org.uk and www.makingthemodern-world.org.uk. He has written and edited nine volumes, including his own *The Uses of Life: A History of Biotechnology* (Cambridge University Press, 1994) and the edited volumes, *Manifesting Medicine* and the Bunge-award winning Garland Encyclopedia *Instruments of Science* which he coedited with Deborah Warner. His history of penicillin, *Penicillin: Triumph and Tragedy*, will be published by Oxford University Press in 2007.

JENNIFER R. FISHMAN is Assistant Professor of Bioethics and Sociology at Case Western Reserve University. She received her Ph.D. in sociology from the University of California, San Francisco. She is currently working on a book analyzing the emergence of Viagra and of researchers' failed attempts to transform Viagra into a drug that might work for women. She continues to track new drug developments in sexual dysfunction research, including investigation of new testosterone therapies for men and women.

JEREMY A. GREENE received both an M.D. and a Ph.D. from Harvard University in the History of Science in June 2005; his dissertation, "The Therapeutic Transition: Pharmaceuticals and the Marketing of Chronic Disease, 1950–2000," narrates the role of pharmaceuticals in the wide-spread treatment of "risk factors" in the second half of the twentieth century. Greene was awarded the American Association for the History of Medicine's 2004 Shryock Medal for his essay, "Releasing the Flood Waters: Diuril and the Reshaping of Hypertension," and the Society for the Social History of Medicine's 2002 Roy Porter Prize for his essay, "Therapeutic Infidelities: 'Noncompliance' Enters the Medical Literature, 1955–1975."

DAVID HEALY studied in University College Dublin and University of Cambridge. He is a Professor in Psychological Medicine at Cardiff University, a former Secretary of the British Association for Psychopharmacology, and author of over 130 peer-reviewed articles and 15 books, including *The Antidepressant Era,* and *The Creation of Psychopharmacology* from Harvard University Press, *The Psychopharmacologists Volumes 1–3,* and *Let Them Eat Prozac* from New York University Press. Healy is also a Visiting Professor at the University of Toronto.

SUZANNE WHITE JUNOD is Senior Historian at the U.S. Food and Drug Administration. She received her Ph.D. from Emory University, has published widely, and is currently writing a book on food safety in America.

ILINA SINGH is a Lecturer in Social Psychology and Research Methodology at the London School of Economics. She works broadly in the developmental psychiatry arena, investigating the psycho-social and ethical implications of neuro- and genetic interventions for children and their families. She has published widely on issues related to ADHD, Ritalin, and other stimulant drugs.

ANDREA TONE is Canada Research Chair in the Social History of Medicine in the Faculties of Medicine and Arts at McGill University. She is the author of numerous articles and books including, most recently, *Devices and Desires: A History of Contraceptives in America.* She is currently writing a history of anxiety and tranquilizers.

ELIZABETH SIEGEL WATKINS is Associate Professor of the History of Health Sciences in the Department of Anthropology, History, and Social Medicine at the University of California, San Francisco. She received her Ph.D. in the History of Science at Harvard University. Watkins is the author of *On the Pill: A Social History of Oral Contraceptives, 1950–1970;* her forthcoming book on the history of hormone replacement therapy will be published by Johns Hopkins University Press in 2007.

Index

"All Are Candidates for
Ritalin" *AJP*, 1968

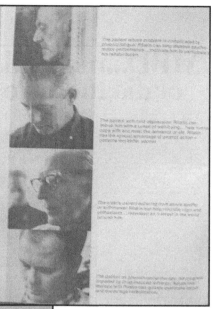

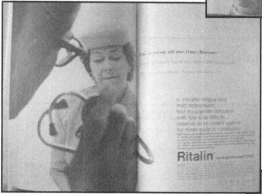

"So Tired" *AJP*, 1964

"Safe to Play" *AJP*, 1958

"Myth or Disease?" *AJP*, 1971

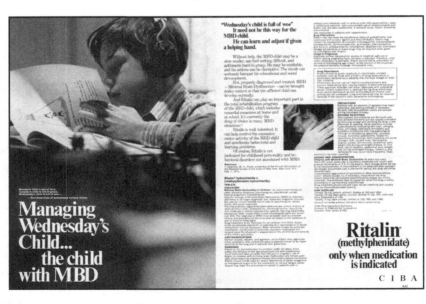

"Wednesday's Child" *AJP*, 1973

"The Coat in the Room," *AJP*, 1974

"Perfect Pictures" *AJP*, 1975

"Domestic Harmony" *People*, 2002

"Finally a Hug"
Good Housekeeping, 2004